THE WORLD BOOK ENCYCLOPEDIA OF
PEOPLE AND PLACES

THE WORLD BOOK ENCYCLOPEDIA OF
PEOPLE AND PLACES

3/I-L

WORLD BOOK, INC.
a Scott Fetzer company
CHICAGO

Acknowledgments:

Front cover
large photo, © Victor Englebert
left inset, © Corbis
center inset, © Tibor Bognar, Corbis Stock Market
right inset, © R & S Michaud, Woodfin Camp, Inc.

Back cover
© Victor Englebert

For information on other World Book publications, visit our Web site at **http://www.worldbook.com** or call **1-800-WORLDBOOK (967-5325).** For information on sales to schools and libraries, call **1-800-975-3250 (United States); 1-800-837-5365 (Canada).**

THE WORLD BOOK ENCYCLOPEDIA OF PEOPLE AND PLACES

This edition published by
World Book, Inc.
233 N. Michigan Ave.
Chicago, IL 60601

The Library of Congress has cataloged a previous edition of this title as follows:

Library of Congress Cataloging-in-Publication Data

The World Book encyclopedia of people and places.
     p. cm.
  Includes index.
  ISBN 0-7166-3750-2
   1. Encyclopedias and dictionaries. [1. Geography—Encyclopedias.] I. World Book, Inc.

AE5 . W563 2001
031.02—dc21

                     2001023546

ISBN 0-7166-3754-5 (this edition)

Printed in the United States of America

15 16 05 04 03

Contents

Political World Map

The world has 193 independent countries and about 40 dependencies. An independent country controls its own affairs. Dependencies are controlled in some way by independent countries. In most cases, an independent country is responsible for the dependency's foreign relations and defense, and some of the dependency's local affairs. However, many dependencies have complete control of their local affairs.

By 2000, the world's population surpassed 6 billion, and the yearly rate of population growth was about 1.4 per cent. At that rate, the world's population would double in about 49 years. Almost all of the world's people live in independent countries. Only about 10 million people live in dependencies.

Some regions of the world, including Antarctica and certain desert areas, have no permanent population. The most densely populated regions of the world are in Europe and in southern and eastern Asia. The world's largest country in terms of population is China, which has more than a billion people. The independent country with the smallest population is Vatican City, with only about 1,000 people. The Vatican City, covering only 1/6 square mile (0.4 square kilometer), is also the smallest in terms of size. The world's largest nation in terms of area is Russia, which covers 6,592,850 square miles (17,075,400 square kilometers).

Every nation depends on other nations in some ways. The interdependence of the entire world and its peoples is called *globalism*. Nations trade with one another to earn money and to obtain manufactured goods or the natural resources that they lack. Nations with similar interests and political beliefs may pledge to support one another in case of war. Developed countries provide developing nations with financial aid and technical assistance. Such aid strengthens trade as well as defense ties.

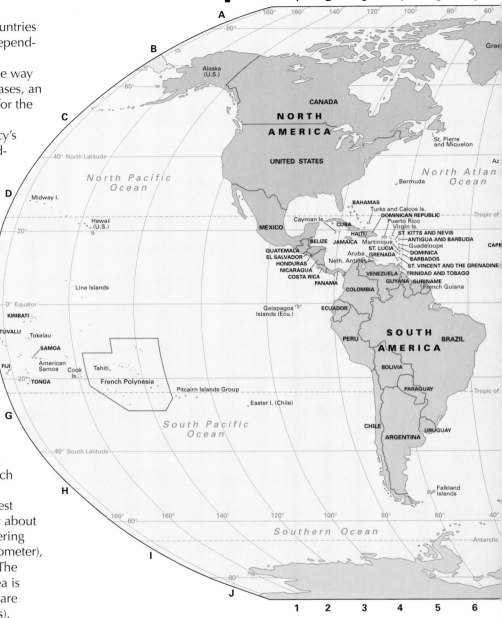

Nations of the World

Name	Map key	Name	Map key	Name	Map key
Afghanistan	D 13	Bulgaria	C 11	Dominican Republic	E 6
Albania	C 11	Burkina Faso	E 9	East Timor	F 16
Algeria	D 10	Burundi	F 11	Ecuador	F 6
Andorra	C 10‡	Cambodia	E 15	Egypt	D 11
Angola	F 10	Cameroon	E 10	El Salvador	E 5
Antigua and Barbuda	E 6	Canada	C 4	Equatorial Guinea	E 10
Argentina	G 6	Cape Verde	E 8	Eritrea	E 11
Armenia	C 12	Central African Republic	E 10	Estonia	E 11
Australia	G 16	Chad	E 10	Ethiopia	E 11
Austria	C 10	Chile	G 6	Federated States of Micronesia	E 18
Azerbaijan	C 12	China	D 14	Fiji	F 1
Bahamas	D 6	Colombia	E 6	Finland	B 11
Bahrain	D 12	Comoros	F 12	France	C 10
Bangladesh	D 14	Congo (Brazzaville)	F 10	Gabon	F 10
Barbados	E 7	Congo (Kinshasa)	F 11	Gambia	E 9
Belarus	C 11	Costa Rica	E 5	Georgia	C 12
Belgium	C 10	Côte d'Ivoire	E 9	Germany	C 10
Belize	E 5	Croatia	C 11	Ghana	E 9
Benin	E 10	Cuba	D 5	Great Britain	C 9
Bhutan	D 14	Cyprus	D 11	Greece	D 11
Bolivia	F 6	Czech Republic	C 11	Grenada	E 6
Bosnia-Herzegovina	C 11	Denmark	C 10	Guatemala	E 5
Botswana	G 11	Djibouti	E 12	Guinea	E 9
Brazil	F 7	Dominica	E 6	Guinea-Bissau	E 9
Brunei	E 15				

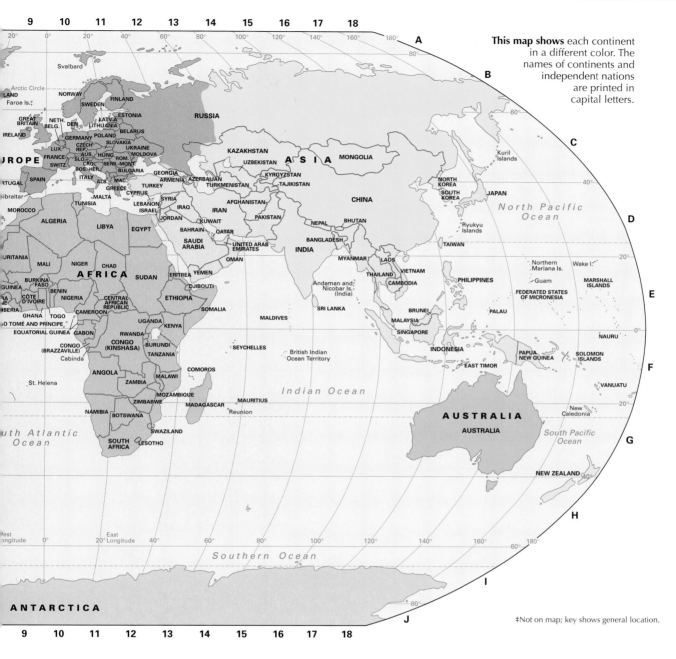

This map shows each continent in a different color. The names of continents and independent nations are printed in capital letters.

‡Not on map; key shows general location.

Iceland

An island country in the North Atlantic Ocean, Iceland lies on the edge of the Arctic Circle between Greenland and Norway. Many people call Iceland the *Land of Frost and Fire* because of its unusual landscape, where large glaciers lie next to steaming hot springs, geysers, and volcanoes. With only 7 persons per square mile (3 persons per square kilometer), this remote island is the most sparsely populated country in Europe.

Iceland's first settlers arrived about 870 with the Norwegian adventurer Ingólfur Arnason. Later, Viking colonists from Britain, some of whom had married Celtic people, also migrated to the island. Present-day Icelanders resemble the people of northern Norway, Ireland, and northern Scotland. About 90 per cent of Iceland's people live close to the coast, which is warmed by the Gulf Stream.

Fishing and fish processing—by far the most important industries in Iceland—provide employment for about 20 per cent of the people, who enjoy a high standard of living. Icelanders spend a large portion of their incomes on their homes, equipping them with television sets, refrigerators, and other

FACT BOX

ICELAND

COUNTRY

Official name: Lyoveldio Island (Republic of Iceland)
Capital: Reykjavik
Terrain: Mostly plateau interspersed with mountain peaks, icefields; coast deeply indented by bays and fiords
Area: 39,769 sq. mi. (103,000 km²)

Climate: Temperate; moderated by North Atlantic Current; mild, windy winters; damp, cool summers
Main rivers: Jökulsa à Fjöllum, Hvitá, Skálfandafljót, Thjórsá
Highest elevation: Hvannadalshnukur, 6,952 ft. (2,119 m)
Lowest elevation: Atlantic Ocean, sea level

GOVERNMENT

Form of government: Constitutional republic
Head of state: President
Head of government: Prime minister
Administrative areas: 23 syslar (counties), 14 kaupstadhir (independent towns)

Legislature: Althing (Parliament) with 63 members serving four-year terms
Court system: Haestirettur (Supreme Court)
Armed forces: The U.S. provides Iceland's defense

PEOPLE

Estimated 2002 population: 286,000
Population growth: 0.57%
Population density: 7 persons per sq. mi. (3 per km²)
Population distribution: 92% urban, 8% rural
Life expectancy in years:
Male: 77
Female: 82
Doctors per 1,000 people: N/A
Percentage of age-appropriate population enrolled in the following educational levels:
Primary: N/A
Secondary: N/A
Further: N/A

modern appliances. They also dress much like people in the United States and Canada. Since much of what they buy must be imported, the cost of living is quite high. As a result, most women work outside the home, and many Icelanders hold more than one job.

Despite their modern lifestyle, Icelanders have kept many connections to the past. Their *Icelandic* language, which is closely related to the Viking tongue of Old Norse, has changed so little through the centuries that people today can easily read tales and poems written in the 1100's and 1200's. In the tradition of their ancestors, Icelanders do not have family names. Instead, they take on a second name by simply adding *-son* (son) or *-dóttir* (daughter) to their father's first name.

A peace-loving and friendly people, today's Icelanders share a long, eventful history trou-bled by natural disasters and political unrest. In the 1200's, civil wars between the early settlers ended only when the *Althing* (parliament) agreed to accept the king of Norway as the ruler of Iceland.

When Norway united with Denmark in 1380, Iceland came under Danish rule. In 1402, the *Black Death* (bubonic plague) swept across the island, and in the late 1700's, volcanic eruptions destroyed livestock and huge areas of farmland, causing massive starvation. During the Napoleonic Wars of the early 1800's, ships bringing food could not reach Iceland, and many people starved to death.

Iceland has been a republic since gaining its independence in 1944. During the late 1980's, the country's fish catch fell, and fish prices dropped on the international market. These factors, along with a worldwide ban on whale hunting, damaged the nation's economy. To reduce its dependence on imports, Iceland has made efforts to expand its range of manufactured products and to grow fruits and vegetables in greenhouses heated by natural hot springs.

Reykjavík, the capital and largest city of Iceland, is located on the southwest coast of the island. All the city's buildings are heated by water piped in from nearby hot springs by the flow of gravity.

Iceland officially gained its independence from Denmark on June 17, 1944. The proclamation was signed at the historic site of Thingvellir, where the Althing—the world's first parliament—was established in 930.

Economy

Language spoken: Icelandic

Religions: Evangelical Lutheran 91% Other Protestant Roman Catholic

Technology

Radios per 1,000 people: N/A

Televisions per 1,000 people: N/A

Computers per 1,000 people: N/A

Currency: Icelandic krona

Gross national income (GNI) in 2000: $8,540 U.S.

Real annual growth rate (1999–2000): 5.0%

GNI per capita (2000): $30,390 U.S.

Balance of payments (2000): N/A

Goods exported: Mostly: Fish and fish products Also: animal products, aluminum, diatomite and ferrosilicon

Goods imported: Machinery and equipment, petroleum products; foodstuffs, textiles

Trading partners: European Union, United States, Japan

Environment

Iceland's natural wonders range from steaming geysers and hot springs to glistening lava fields and fiery volcanoes. This unusual landscape has been shaped by the *fault line* that runs across the island.

The fault line represents a boundary between two gigantic *plates* in the earth's crust—the North American Plate and the Eurasian Plate. These plates move about 1/2 inch to 4 inches (1.3 to 10 centimeters) a year, and as they slide past one another, the motion produces a strain on the rocks on either side of the fault. The strain of the motion often causes earthquakes.

Iceland was formed millions of years ago, when undersea lava flowing from a ridge on the seabed reached the surface of the sea. As recently as 1963, a volcano erupted south of Iceland, forming a new island called Surtsey. Iceland itself has about 200 volcanoes, and their eruptions have spread lava and volcanic rock over the land throughout the island's history.

Volcanoes and geysers

Iceland's most famous volcano is Mount Hekla, which rises 4,892 feet (1,491 meters) above sea level. There have been 18 eruptions on or near Hekla since the 1100's. The main crater of the volcano had been quiet for more than 100 years when it suddenly erupted in 1947. The eruption lasted for 13 months, spreading fiery lava over an area of 15 square miles (39 square kilometers).

Accompanying this volcanic activity are the geysers—explosions of hot water, cloudy with steam, which burst forth from the ground in huge columns. The word *geyser,* in fact, comes from the name of Iceland's most famous hot spring, *Geysir,* which spouts water about 195 feet (59 meters) into the air. Only 70 miles (110 kilometers) from the capital city of Reykjavík, dozens of geysers appear within a circle of 10 miles (16 kilometers).

Glaciers cover about one-eighth of Iceland's surface. The largest glacier, Vatnajökull, covers 3,140 square miles (8,133 square kilometers) and is as big as all the glaciers on the European continent combined.

Green fields surround a small farm in the coastal lowlands of Iceland, *below.* In addition to cattle, most farmers raise sheep for their meat and wool. Icelanders eat more lamb and fish than do people in most other countries.

These mighty glaciers have left their mark on the Icelandic landscape by carving deep trenches in the bottoms of many fiords, creating natural harbors. The holes made by glaciers on land have made numerous lakes, while water from melting glaciers forms rushing rivers and spectacular waterfalls. The most beautiful waterfalls in Iceland are Gullfoss in the south and Dettifoss in the north.

Climate and land regions

The warm Gulf Stream current reaches Iceland from the south, warming the coastal lowlands throughout the year and keeping the harbors free of ice. As a result, Iceland enjoys a comparatively mild climate, with frequent rainfall, patchy sunlight, and medium-force winds. Summers are mild and winters are cool.

Iceland's many volcanoes have formed a huge inland plateau, which covers most of the interior of the island. Here, the land surface is made up of *basalt* (hard volcanic

High pastures fringe an icefield on the edge of Iceland's inland plateau. The lava flow in the foreground is a reminder of the island's volcanic origins. Iceland has about 200 active volcanoes, and there are also active volcanoes under the sea off the coasts.

rock), and lumps of volcanic lava create a barren, almost unearthly, landscape. Little vegetation can take root in this region because the porous volcanic surface discourages plant growth.

The coastal areas, by contrast, contain fertile farmland. The chief crop is hay, which is used to feed sheep, cattle, and the small Icelandic horses raised on the island's 5,300 farms. Heavy rainfall and the long hours of summer sunshine caused by the midnight sun allow Icelandic farmers to grow two or three crops of hay each year. Some farmers also grow root crops, such as turnips and potatoes.

After lying dormant for more than 5,000 years, a volcano on the island of Heimaey, *left,* off Iceland's southern coast, erupted in bursts of fiery lava in 1973. The volcano poured ash over Heimaey's only town, Vestmannaeyjar, forcing the evacuation of all the residents. The lava flow enlarged the island by almost 1 square mile (2.6 square kilometers), and after the lava had cooled and solidified, it was found to have improved the shape of the island's main harbor. The harbor's entrance was narrowed, thus providing greater protection to boats within the harbor.

In southwestern Iceland, the Svartsengi geothermal power station, *below left,* taps energy from deep within the earth's crust. Many people bathe in the warm blue waters nearby, believing that dissolved minerals help treat skin conditions and breathing problems.

Fish are processed in a factory near Reykjavík. Fishing crews in large *trawlers* (fishing boats) drag nets along the seabed to catch capelin, cod, haddock, herring, and flounder. Inland lakes and rivers contain plentiful supplies of trout and salmon.

Icelandic Sagas

According to Norse mythology, two places existed before the creation of life—Muspellsheim, a land of fire, and Niflheim, a land of ice and mist. Between them lay Ginnungagap, a great emptiness where heat and ice met. Where Muspellsheim and Niflheim met, fire thawed ice to set the stage for the creation of the world.

The Viking seafarers who discovered Iceland about 870 must have thought the old legend had come to life when they saw its strange landscape. Towering glaciers and bleak stone deserts, like those of Niflheim, lay alongside active volcanoes and fiery lava, like those of Muspellsheim.

Ancient Viking tales

These Norse adventurers brought with them a wealth of Scandinavian tradition, including poems and stories that were told and retold around roaring fires during the long, dark nights of winter. Gradually, these age-old tales were carefully written down, then copied and recopied through the years.

These stories and poems gave rise to *sagas,* a great body of literature written in Iceland between the 1100's and the 1300's. The word *saga* is related to the Icelandic verb ''to say'' or ''to tell.'' Beginning in the 1100's, Icelanders such as Saemund the Learned, Ari Thorgilsson, and Snorri Sturluson set down in writing the tales of their Scandinavian ancestors, as well as detailed records of their own Icelandic history.

The great poet and historian Snorri Sturluson wrote *Heimskringla (Circle of the World)*—a history of the kings of Norway from their origins up to his own day. He probably also wrote *The Saga of Egill Skallagrimsson,* one of the best Icelandic sagas, about a great poet of the 900's who was one of Snorri's forefathers.

Snorri also composed the *Prose Edda,* whose first section narrates myths about Scandinavian gods. Two centuries after Iceland officially converted to Christianity, Snorri Sturluson gave his people an extraordinary collection of the beliefs of his

A cloud of steam drifting across the barren volcanic landscape of Iceland—which is virtually unchanged since the sagas were written—brings to mind the Norse creation myth, which speaks of a place where the land of frost meets the land of fire.

A statue of Leif Ericson, an early Norse explorer whose life was recorded in the Icelandic sagas, looks out over the hills of Reykjavík.

Fish are dried on this large framework, *above,* before being preserved in salt. The Norse seafarers, whose adventures live on through the sagas, depended heavily on preserved fish during their long voyages.

The Vikings were fierce pirates and daring sailors who explored the North Atlantic Ocean from the late 700's to about 1100. Vikings were among the best shipbuilders of their time, constructing vessels for trade and warfare from wood cut from the plentiful Scandinavian forests. Their trading ships, called *knorrs,* were about 50 feet (15 meters) long.

ancestors, leaving a dramatic picture of Viking gods and heroes for generations to come.

The classic stories

The classic stories, known in English as *Icelandic Family Sagas,* were composed in the 1200's by anonymous writers. These sagas vary in length from brief stories to the equivalent of full-length novels. They relate the adventures of Icelandic and Scandinavian heroes—famous people of the past who had a passion for honor, a respect for poetry and learning, and a fierce desire for independence.

The sagas bring to vivid life the people who gave Iceland its dramatic and eventful past. Icelanders still enjoy the tale of Helgi the Lean, who "believed in Christ, but prayed to Thor [a Norse god] for sea voy-

ages and in times of danger." And people today are still inspired by the story of Njal Thorgeirsson, who strove for peace in a violent world, achieving it only by sacrificing himself and his family.

Scholars once thought that the classic sagas had been told and retold for generations before being written down during the 1200's. Today, however, most experts believe the sagas were artistic creations based on oral and written traditions. They were written during a period of moral and social decline and upheld the high values of a previous "golden age" that occurred between 850 and 1050.

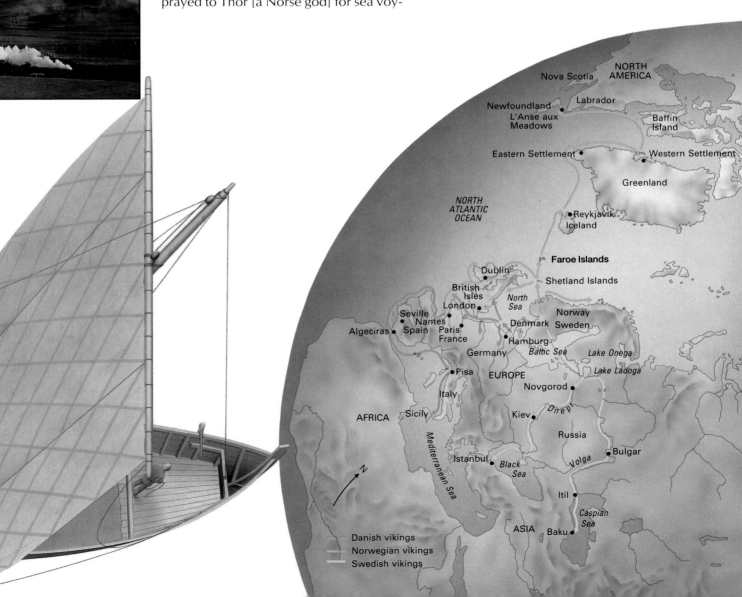

India

India, the seventh largest country in the world in area, is a huge peninsula, extending far out into the Indian Ocean. High mountain ranges extend along India's northern border and separate most of the country from the rest of Asia. Pakistan lies on India's northwest border. To the northeast, India borders China, Nepal, Bhutan, Myanmar, and Bangladesh.

Although India covers less than 2-1/2 per cent of the world's land area, it has about 16 per cent of the world's population. Only China has a greater number of people than India. The river valleys of northern India are among the most populated places in the world. Most of India's people practice the religion of Hinduism.

For centuries, the Western world looked upon India as a land of mystery, wealth, and excitement. The glories of its past can be seen in the Taj Mahal in northern India. This splendid white marble structure was built by a Mogul ruler between about 1630 and 1650 as a tomb for his favorite wife. The tomb stands at Agra in northern India. About 20,000 workers built the structure, which, according to tradition, was designed by a Turkish architect. It is still one of the world's most famous and beautiful buildings.

Today, India is a fascinating mixture of old and new. Traffic on busy city streets is often stopped by wandering cattle, which Hindus consider sacred and allow to roam freely. Many Indians go to work in modern factories wearing traditional costumes and carrying supplies on their heads in handwoven baskets.

India is a scenic land of dramatic contrasts and variety. Along its northern border lies the Himalaya, the tallest mountain system in the world, while jungles on the southern peninsula are the natural habitat of elephants and other large animals. The northern plains include the world's largest *alluvial plain* (land formed of soil left by rivers).

The people of India belong to many different ethnic groups and their way of life varies widely from region to region. They speak 16 major languages and more than 1,000 minor languages and dialects. Some Indians are very wealthy, but many others are so poor that they have no homes and must sleep in the streets. Some Indians are college graduates, but many others have never gone to school at all.

India Today

A British colony since the late 1700's, India won its independence on Aug. 15, 1947.

During India's struggle for freedom, Hindus and Muslims fought to gain political power for their own group. In an effort to stop the fighting, Indian and British leaders divided the country into two nations: India, for the Hindus; and Pakistan, for the Muslims. The division led to even more fighting.

The area of Kashmir, on the northern border of India, has been the focus of the struggle between Muslims and Hindus since 1947. When India was divided into India and Pakistan, each new country claimed Kashmir as its own. Kashmir's ruler was Hindu, but most of its people were Muslims. When Pakistan invaded the area, Kashmir's ruler made it part of India for protection. War raged on until 1949, when the United Nations arranged a cease-fire.

Another border dispute, this time with China, turned into armed conflict in 1959. Then, in 1962, China invaded India. By the end of that year, China had pulled back, and a cease-fire took effect.

In the 1980's, followers of the religion of Sikhism demanded more political control of their home state of Punjab. Some Sikhs carried out terrorist acts, but these lessened in the 1990's.

In 1984, India's prime minister Indira Gandhi was assassinated by two Sikh members of her security force. After her assassination, the Congress-I party chose Rajiv Gandhi—her son—as prime minister. Rajiv Gandhi remained prime minister until 1989, when the party lost its majority in Parliament. Rajiv Gandhi was assassinated in 1991 while campaigning for the office of prime minister.

Several Indian regions suffered violence in the 1990's. In the state of Jammu and Kashmir, war broke out between security forces and Muslim guerrillas seeking independence. India was accused of violating human rights after it assumed emergency powers. In 1994, India canceled elections scheduled for the state and barred human rights groups from Kashmir. Also in 1994, farmers belonging to the Bodo tribe in Assam state attacked newcomers settling on local farmland. About 100 people died.

In September 1993, an earthquake struck a rural area southeast of Bombay, now called Mumbai, killing about 10,000 people. Another devastating earthquake hit India in January 2001. The quake, which was centered in the far west of India near the city of Bhuj, cost tens of thousands of people their lives and caused billions of dollars of damage.

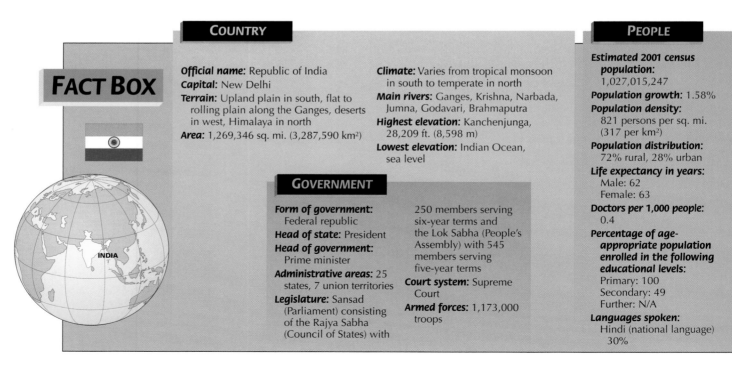

FACT BOX

COUNTRY

Official name: Republic of India
Capital: New Delhi
Terrain: Upland plain in south, flat to rolling plain along the Ganges, deserts in west, Himalaya in north
Area: 1,269,346 sq. mi. (3,287,590 km²)

Climate: Varies from tropical monsoon in south to temperate in north
Main rivers: Ganges, Krishna, Narbada, Jumna, Godavari, Brahmaputra
Highest elevation: Kanchenjunga, 28,209 ft. (8,598 m)
Lowest elevation: Indian Ocean, sea level

GOVERNMENT

Form of government: Federal republic
Head of state: President
Head of government: Prime minister
Administrative areas: 25 states, 7 union territories
Legislature: Sansad (Parliament) consisting of the Rajya Sabha (Council of States) with 250 members serving six-year terms and the Lok Sabha (People's Assembly) with 545 members serving five-year terms
Court system: Supreme Court
Armed forces: 1,173,000 troops

PEOPLE

Estimated 2001 census population: 1,027,015,247
Population growth: 1.58%
Population density: 821 persons per sq. mi. (317 per km²)
Population distribution: 72% rural, 28% urban
Life expectancy in years:
Male: 62
Female: 63
Doctors per 1,000 people: 0.4
Percentage of age-appropriate population enrolled in the following educational levels:
Primary: 100
Secondary: 49
Further: N/A
Languages spoken: Hindi (national language) 30%

INDIA

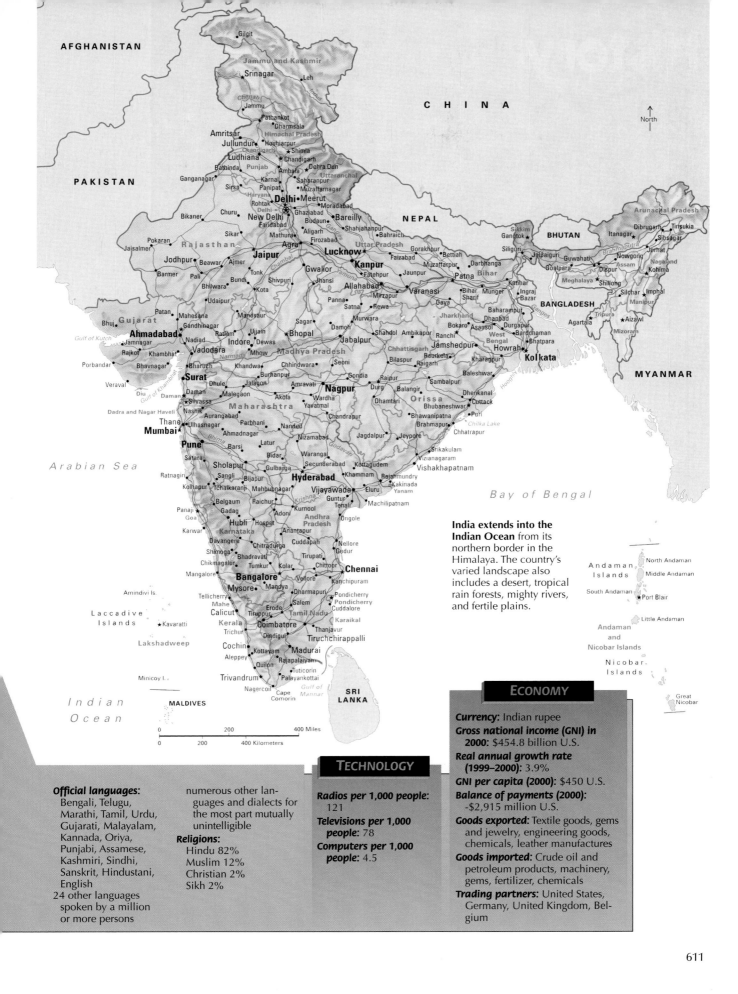

AFGHANISTAN

CHINA

PAKISTAN

Gilgit

Jammu and Kashmir
•Srinagar   •Leh

•Jammu

•Amritsar   •Pathankot   •Dharmsala
Jullundur•   •Hoshiarpur   **Himachal Pradesh**
•Ludhiana   ★Shimla
Bathinda•   Chandigarh★   Dehra Dun•
Ganganagar•   Karnal•   Saharanpur•   **Uttaranchal**
•Sirsa   Panipat•   Muzaffarnagar•
•Rohtak •Meerut
New Delhi   •Moradabad
•Faridabad   •Ghaziabad   •Budaun   •Bareilly   **NEPAL**
•Bikaner   •Churu   Mathura•   Shahjahanpur•   •Bahraich
Pokaran•   •Sikar   Aligarh•   •Lucknow   Gorakhpur•
Jaisalmer•   **Rajasthan**   •Agra   **Uttar Pradesh**   •Faizabad   Bettiah•
•Jodhpur   Beawar• •Ajmer   •Jaipur   •Gwalior   •Kanpur   Jaunpur•   Muzaffarpur•
•Barmer   •Pali   •Tonk   Shivpuri•   Jhansi•   •Fatehpur   Darbhanga•
Bhilwara•   •Bundi   •Kota   •Allahabad   •Varanasi   **Bihar**   Katihar•
•Udaipur   Panna•   Mirzapur•   Bihar   •Munger   Ingraj
•Patan   Mandsaur•   •Rewa   Gaya   Sharif•   Bazar•
Bhuj• **Gujarat**   •Mahesana   •Ratlam   •Ujjain   •Sagar   •Satna   Murwara•   **Jharkhand**   Dhanbad•
Jamnagar• Gandhinagar•   •Indore   Dewas•   •Damoh   •Shahdol   Ambikapur•   Bokaro• •Asansol
•Rajkot •Nadiad •Mhow   **Madhya Pradesh**   •Jabalpur   •Ranchi   Durgapur•
Porbandar• •Khambhat   Khandwa•   •Bhopal   Chhattisgarh   •Jamshedpur
•Bharuch   *Narmada*   •Chhindwara   Bilaspur•   Raurkela•   Barddhaman•   Bhatpara•
Bhavnagar•   Dhule•   •Seoni   •Raigarh   **West**   •Kharagpur   **Howrah**
•Veraval   •Surat   Jalgaon•   Gondia•   •Raipur   **Bengal**   Baleshwar•
•Diu   Daman•   Malegaon•   Amravati•   •Nagpur   •Durg   •Balangir   •Sambalpur   **Kolkata**
★Silvassa   •Nashik   Akola•   •Wardha   Dhamtari•   Dhenkanal•
Dadra and Nagar Haveli   •Aurangabad   •Yavatmal   **Orissa**   •Cuttack
•Thane   Ulhasnagar•   Parbhani•   •Nanded   Chandrapur•   Bhawanipatna•   •Bhubaneswar
Mumbai   Ahmadnagar•   Nizamabad•   Jagdalpur•   Brahmapur•   •Puri
Pune   Barsi•   **Maharashtra**   •Latur   •Warangal   Jeypore•   Chhatrapur•   *Chilka Lake*
•Satara   •Bidar   Secunderabad•   Srikakulam•
Ratnagiri•   Sholapur•   •Gulbarga   **Hyderabad**   Kottagudem•   Vizianagaram•
Sangli•   Bijapur•   Khammam•   •Vishakhapatnam
•Kolhapur   Ichalkaranji•   Mahbubnagar•   •Vijayawada   Rajahmundry•
•Belgaum   •Raichur   Eluru•   Kakinada•
•Panaji   •Gadag   Kurnool•   •Guntur   Yanam•
Goa   •Hubli   Adoni•   **Andhra**   Tenali•   Machilipatnam•
•Karwar   •Hospet   Anantapur•   **Pradesh**   Ongole•
Karnataka   •Chitradurga   Cuddapah•
•Davangere   •Nellore
Shimoga•   Bhadravati•   Tirupati•   •Gudur
•Mangalore   Chikmagalur•   •Tumkur   •Kolar   •Chittoor
Bangalore   •Vellore   Kanchipuram•   **Chennai**
•Mysore   Mandya•   Dharmapuri•
Tellicherry•   •Salem   •Pondicherry
•Mahe   Erode•   •Cuddalore
•Calicut   •Tiruppur   **Tamil Nadu**
•Trichur   •Coimbatore   •Karaikal
Kerala   •Dindigul   •Thanjavur
•Cochin   •Tiruchchirappalli
Aleppey• •Kottayam   •Madurai
•Quilon   Rajapalaiyam•
•Tuticorin
Trivandrum★   Palayankottai•
•Nagercoil   Cape Comorin

Arabian Sea

North

Arunachal Pradesh
•Dibrugarh •Tinsukia
Itanagar•   •Sibsagar
Sikkim   Nowgong• •Jorhat
Gangtok★   **BHUTAN**   Jalpaiguri•   Nagaland
Siliguri•   •Guwahati   **Assam**   •Dispur
Meghalaya   •Shillong   •Kohima
Goalpara•
BANGLADESH   **Tripura**   •Aizawl
Baharampur•   Agartala•   •Imphal
•Silchar   **Manipur**
Ganges   **Mizoram**

MYANMAR

Hooghly

Bhima   *Krishna*   *Godavari*

Bay of Bengal

India extends into the Indian Ocean from its northern border in the Himalaya. The country's varied landscape also includes a desert, tropical rain forests, mighty rivers, and fertile plains.

Amindivi Is.

Laccadive Islands

Lakshadweep   ★Kavaratti

Minicoy I.•

Indian Ocean

MALDIVES

Gulf of Mannar
Cape Comorin

SRI LANKA

Andaman Islands   •North Andaman
  •Middle Andaman
South Andaman   ★Port Blair
  •Little Andaman
Andaman and Nicobar Islands
Nicobar Islands
  •Great Nicobar

MALES

0	200 400 Miles
0	200 400 Kilometers

Gulf of Kutch
Gulf of Khambhat

History

About 1500 B.C., a tribe of fair-skinned people called *Aryans* migrated to India from central Asia. They are the ancestors of present-day northern Indians. When the Aryans arrived, the dark-skinned Dravidians, who were already living in India, fled to the south. They are the ancestors of the people who live in southern India today.

The influence of the Aryans gradually spread through all of India. The Aryans developed the Sanskrit language and established a way of writing it. Also during the time of Aryan rule, Siddhartha Gautama (563?-483? B.C.) founded the religion of Buddhism.

Empires and conquerors

In 326 B.C., Alexander the Great conquered the northwestern region of India. After his death, Chandragupta Maurya founded the Maurya Empire, the first to unite almost all of India under one government. The most famous Maurya ruler, Emperor Asoka, helped spread Buddhism throughout the country.

About A.D. 120, the Maurya Empire fell to the Scythians, who ruled northern India as the Kushan dynasty. Later, native Indian emperors of the Gupta dynasty extended their empire across all of northern India and parts of Afghanistan.

The golden age of India

The Gupta Empire, which lasted from 320 to about 500, was the golden age of India. During this period, the arts flourished. Sanskrit, the classical language of Hindu religion and culture, flowered in poetry and literature.

The prosperous upper classes of Gupta cities supported music, dance, and other cultural activities while Hindu and Buddhist sculpture became the model for later Indian art. Hindu schools taught grammar, mathematics, medicine, philosophy, and sacred writings. Students came from as far away as China and Java to study at Gupta universities.

After the fall of the Gupta Empire, India was invaded by the Huns. In 1526, the

Jawaharlal Nehru, *above,* served as the first prime minister of independent India from 1947 until his death in 1964. A follower of Gandhi, Nehru worked to establish a democracy and to improve the living standards of the Indian people. His daughter, Indira Gandhi, and his grandson, Rajiv Gandhi, have also served as prime ministers.

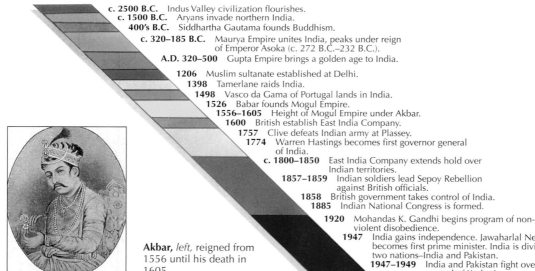

c. 2500 B.C.   Indus Valley civilization flourishes.
c. 1500 B.C.   Aryans invade northern India.
400's B.C.   Siddhartha Gautama founds Buddhism.
c. 320–185 B.C.   Maurya Empire unites India, peaks under reign of Emperor Asoka (c. 272 B.C.–232 B.C.).
A.D. 320–500   Gupta Empire brings a golden age to India.
1206   Muslim sultanate established at Delhi.
1398   Tamerlane raids India.
1498   Vasco da Gama of Portugal lands in India.
1526   Babar founds Mogul Empire.
1556–1605   Height of Mogul Empire under Akbar.
1600   British establish East India Company.
1757   Clive defeats Indian army at Plassey.
1774   Warren Hastings becomes first governor general of India.
c. 1800–1850   East India Company extends hold over Indian territories.
1857–1859   Indian soldiers lead Sepoy Rebellion against British officials.
1858   British government takes control of India.
1885   Indian National Congress is formed.
1920   Mohandas K. Gandhi begins program of non-violent disobedience.
1947   India gains independence. Jawaharlal Nehru becomes first prime minister. India is divided into two nations–India and Pakistan.
1947–1949   India and Pakistan fight over control of Kashmir.
1948   Gandhi is assassinated.
1950   India becomes an independent, democratic republic.
1962   Chinese forces invade India over a border dispute, but pull back.
1965   India and Pakistan fight second war over control of Kashmir.
1966   Indira Gandhi, daughter of Nehru, becomes prime minister.
1971   East Pakistan becomes an independent nation called Bangladesh. India supports Bangladesh in civil war with West Pakistan.
1989   Congress-I Party is defeated.
1991   Rajiv Gandhi is assassinated.
1999   Heavy fighting takes place between Indian troops and Muslim guerrillas in Kashmir.

Akbar, *left,* reigned from 1556 until his death in 1605.

Mohandas K. Gandhi, *far left,* led the Indian National Congress from 1920 until his death in 1948.

Indira Gandhi, *left,* was prime minister from 1966 until her death in 1984.

Muslim ruler Babar invaded India and established the Mogul Empire. Babar's grandson, Akbar, ruled India wisely until his death in 1605 and won the loyalty of the Hindus. However, a harsh ruler named Aurangzeb, who became emperor in 1658, taxed the Hindus unfairly and destroyed their temples.

British India

With the death of Aurangzeb in 1707, there was no longer a central power in India. The East India Company of England, which had already set up trading posts in Bombay, Calcutta, and Madras (now called Mumbai, Kolkata, and Chennai), took advantage of this opportunity. In the mid-1700's, the company took control of much of India.

In 1857, Indian soldiers, called *sepoys,* rebelled against British officers who ordered them to bite open cartridges greased with fat from cows or hogs. The Hindu soldiers refused to obey the order because their religion did not allow them to eat beef. The Muslim soldiers refused because they were forbidden to eat pork by their religion.

After the defeat of the Sepoy Rebellion, in 1858 the British government took control of all the land that had been governed by the East India Company. The area then became known as *British India.*

During their rule, the British built railroad, telegraph, and telephone systems in India. They also enlarged the Indian irrigation system.

However, the Indian people criticized British policies and accused the British of holding them back in job opportunities and other areas. In 1885, with British approval, the Indian people formed the Indian National Congress to discuss their problems.

In 1920, Mohandas K. Gandhi became the leader of the Indian National Congress. Under his leadership, millions of Indians joined the movement for independence. Gandhi defied British rule with a policy of *nonviolent disobedience,* which included nonpayment of taxes, sit-ins, and refusal to attend British schools. In 1947, the Indian people won their independence from the United Kingdom.

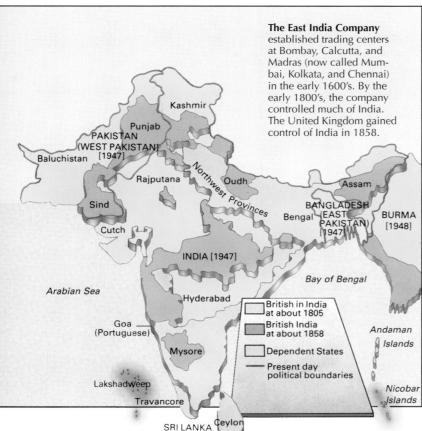

The East India Company established trading centers at Bombay, Calcutta, and Madras (now called Mumbai, Kolkata, and Chennai) in the early 1600's. By the early 1800's, the company controlled much of India. The United Kingdom gained control of India in 1858.

Kashmir

Punjab
PAKISTAN
(WEST PAKISTAN)
[1947]
Baluchistan

Rajputana · Northwest Provinces · Oudh

Assam

Sind

BANGLADESH
(EAST PAKISTAN)
[1947]
Bengal

BURMA
[1948]

Cutch

INDIA [1947]

Arabian Sea

Bay of Bengal

Hyderabad

Goa
(Portuguese)

Andaman Islands

Mysore

British in India at about 1805
British India at about 1858
Dependent States
Present day political boundaries

Lakshadweep

Travancore

Nicobar Islands

SRI LANKA Ceylon
[1948]

Environment

India's land mass is shaped like a triangle with its northern base in the Himalaya. Farther south, India extends into the Indian Ocean. The Arabian Sea borders its western edge, and the Bay of Bengal lies to the east.

India can be divided into three major land regions: (1) the Himalaya, (2) the Northern Plains, and (3) the Deccan, or Southern Plateau.

The Himalaya

The Himalaya mountain system extends along India's northern and northeastern border in a 1,500-mile (2,410-kilometer) curve.

The Himalaya is the tallest mountain system in the world. Kanchenjunga, India's tallest mountain, rises 28,209 feet (8,598 meters) high in the Himalaya on the Nepal border.

Because of the difference in altitude and exposure in many parts of the range, climate and plant life in the Himalaya are quite varied. Tropical plants, such as fig and palm trees, grow on the steep southern slopes up to 3,000 feet (910 meters). At 12,000 feet (3,660 meters), deodar and pine trees flourish.

Many kinds of animals live in the Himalaya, from tigers and leopards to yaks, rhinoceroses, elephants, and some kinds of monkeys.

The Northern Plains

The Northern Plains, which lie between the Himalaya and the southern peninsula, include the valleys of the Brahmaputra, Ganges, and Indus rivers and their branches. The Northern Plains make up the world's largest *alluvial plain* (land formed of soil left by rivers). The soil in this region is excellent farmland, among the most fertile in the world. Irrigation is easy because the land is flat.

The Deccan

The Deccan, a huge plateau that forms most of the southern peninsula, is completely surrounded by mountains. To the north, the Vindhya and other mountain ranges separate the Deccan from the Northern Plains.

To the east, the Eastern Ghats rises along the edge of the Deccan and gradually slopes to a wide coastal plain along the Bay of Bengal. The Western Ghats on the western edge of the Deccan falls sharply to a narrow coastal plain along the Arabian Sea. The Eastern and Western Ghats meet in the south in the Nilgiri Hills.

The Deccan includes farming and grazing land, as well as most of India's valuable mineral deposits. The forests along its eastern and western edges are the natural habitats of many large animals.

Climate

India's climate is quite varied. During the cool season (October through February), the weather is mild, except in the northern mountains where snow falls. The Northern

Village women of the Northern Plains collect the droppings of domestic cattle, which they dry in the sun and use for fuel. Most of India's people live in this fertile area.

Plains get the most intense heat during the hot season, which lasts from March to the end of June. The rainy season lasts from the middle of June through September, as the *monsoons* (seasonal winds) blow across the Indian Ocean from the southwest and southeast.

India's climate can be violent, with heavy rainfall, flash floods, extreme temperatures, and fierce winds that often occur without warning.

India can be divided into three major land regions: (1) the soaring mountains of the Himalaya, (2) the fertile farmland of the Northern Plains, and (3) the Deccan, with its farmland, forests, and rich mineral deposits.

A highway winds through a tiny settlement in Himachal Pradesh toward a high pass in the Himalaya. These mountains separate India from China and the rest of Asia. The name *Himalaya* means *House of Snow,* or the *Snowy Range,* in Sanskrit.

Terraced fields extend down a slope in the Nilgiri Hills of southern India. The Nilgiris are the meeting point of the Eastern and Western Ghats on the Deccan, a vast upland area that covers half of India.

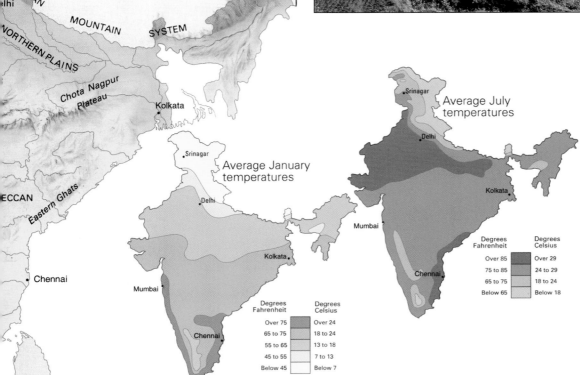

Average July temperatures

Average January temperatures

Degrees Fahrenheit	Degrees Celsius
Over 85	Over 29
75 to 85	24 to 29
65 to 75	18 to 24
Below 65	Below 18

Degrees Fahrenheit	Degrees Celsius
Over 75	Over 24
65 to 75	18 to 24
55 to 65	13 to 18
45 to 55	7 to 13
Below 45	Below 7

Temperature maps for January and July show the wide range in India's weather. The cool winds of the winter monsoon keep temperatures low in the northern areas while in southern India, temperatures stay fairly high. By July, temperatures have already peaked, as the summer monsoon has brought rainfall. The hottest weather occurs between March and June. The Himalaya stays cool throughout the year while the southern coasts have high temperatures all year long.

615

The Monsoons

The *monsoons* are seasonal winds that blow across the northern part of the Indian Ocean, especially the Arabian Sea. They also blow across most of the surrounding land areas.

Monsoons are caused by the difference in the heating and cooling of air over land and sea. India has two monsoon seasons: the moist summer monsoon blows from the southeast and southwest from the middle of June through September, and the dry, cool winter monsoon blows from the northeast from November to March.

The summer monsoon is extremely important to agriculture in India because it brings much-needed rainfall. The winter monsoon brings very little rain because the cool air blowing from the Himalaya does not carry as much moisture as warm air.

Monsoons have an important effect on the nation's agriculture, so Indian meteorologists spend a great deal of time trying to understand and predict the behavior of these winds. Monsoons are known to be influenced by sea temperatures as far away as the South American coast. Other factors, such as a westerly *jet stream* (a fast-moving, high-altitude airstream)—or even the depth of snow on the Himalaya—may also affect the monsoons.

The summer monsoon

During the summer, the sun heats land surfaces far more than it does sea surfaces. As this strongly heated air over the land rises, it is replaced by a southwesterly wind carrying warm, moist air from the Indian Ocean.

The rising air absorbs moisture in the form of water vapor as it heats up. The vapor condenses and forms clouds made up of water droplets, which fall as rain. This process releases large amounts of heat, which helps drive monsoons.

The summer monsoon reaches India from the southeast and southwest from the middle of June through September.

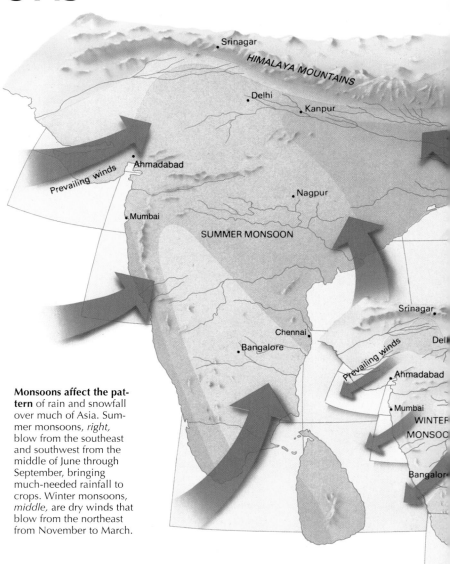

Monsoons affect the pattern of rain and snowfall over much of Asia. Summer monsoons, *right,* blow from the southeast and southwest from the middle of June through September, bringing much-needed rainfall to crops. Winter monsoons, *middle,* are dry winds that blow from the northeast from November to March.

Floods from monsoon rains may claim many lives in addition to destroying crops. In areas especially affected by flooding, houses may be built on stilts, to keep them above the level of the floodwaters.

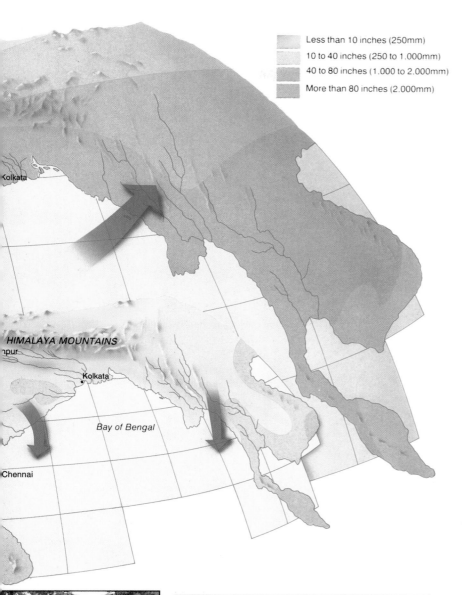

Less than 10 inches (250mm)
10 to 40 inches (250 to 1.000mm)
40 to 80 inches (1.000 to 2.000mm)
More than 80 inches (2.000mm)

Kolkata

HIMALAYA MOUNTAINS

Kolkata

Bay of Bengal

Chennai

Effect on agriculture

The arrival of the summer monsoon has an extremely important effect on Indian agriculture. Farmers depend on the rainfall brought by the summer monsoon to water their crops—if the monsoon brings enough rain, the crops will grow.

However, the monsoon does not always behave as expected. Sometimes the monsoon is late. Sometimes the monsoon rains start and then stop prematurely. And sometimes they turn back over the ocean. Then the crops fail because they do not get enough rainfall.

A poor monsoon season creates other serious problems. Water reserves usually saved for later in the season are used up. Hydroelectricity plants are sometimes closed down and farmers are left without electricity to run their well pumps. And sometimes the monsoon rains are so heavy that they cause serious flooding and crop destruction. Flooding in some areas and drought in other areas can occur in the same monsoon season.

In 1987, large parts of India suffered the worst drought in 40 years, while other areas saw the worst flooding in decades. The northwest state of Rajasthan suffered through a drought, while Assam in the northeast recorded the highest July rainfall in 50 years.

In northern Bihar, tens of thousands of villages were flooded repeatedly from July through the end of the rainy season. During the same period, villagers in other regions lost their crops as the sun scorched their land and wells as deep as 150 feet (46 meters) dried up.

Although the failure of the monsoon in 1987 was far worse than the droughts of 1965 and 1979, the effects were not as severe. Millions of tons of food grains were lost, but stores of cereals from previous plentiful harvests helped replace the lost grain.

Umbrellas often appear on the streets of bustling Mumbai during the summer monsoon. Indian people accept the threat of flooding and crop destruction in return for the life-giving rains of the summer monsoon.

Agriculture

Agriculture is the largest and most important part of India's economy. Farms cover more than half of the country's area and about 60 per cent of Indian workers make their living by farming.

India's farmers grow a variety of crops, and India has the world's largest cattle population. Because cows are sacred to Hindus, they are not butchered for meat. Farmers keep cows for milk production and for plowing. After the cows die, their hides are used to make leather.

Crop production

About 80 per cent of India's farmland is used to grow grains and *pulses,* the country's main foods. Pulses are the seeds of various pod vegetables, such as beans, chickpeas, and pigeon peas. The major grain crops are rice, wheat, millet, and sorghum.

India grows half of the world's mangoes and ranks as the world's leading producer of cashews, millet, peanuts, pulses, sesame seeds, and tea. It is also the world's largest grower of *betel nuts,* which are palm nuts chewed as a stimulant by many people in tropical Asia.

India is one of the world's leading producers of cauliflowers, jute, onions, rice, sorghum, and sugar cane. Indian farmers are also major producers of apples, bananas, coconuts, coffee, cotton, eggplants, oranges, potatoes, rapeseeds, rubber, tobacco, and wheat. In addition, they grow such spices as cardamom, ginger, pepper, and turmeric.

Indian farms

Most farms in India are quite small. Half are less than 2-1/2 acres (1 hectare) in area and only 4 per cent cover more than 25 acres (10 hectares).

About two-thirds of India's farmers own their own land, but many of these farms become smaller with each succeeding generation because of Indian inheritance customs. When a man dies, his land is divided up equally among his sons. In time, the property often becomes too small to cultivate profitably. To solve this problem, many state governments have set limits on how much a farm can be divided.

At harvesttime, women gather in the rice fields to cut the stalks with sickles. Rice is the staple crop in the eastern and southern regions. The world's second largest rice producer, India also exports some high-quality rice.

With the help of a pair of oxen, *right,* a farmer plows a wheat field near one of the famous Hindu temples in Khajuraho. Wheat is one of northern India's main foods. The people enjoy *chapatties,* thin, flat, baked breads made of wheat flour.

India's farmers have two growing seasons: the main summer cultivation period, called the *kharif;* and the secondary winter season, the *rabi.* The kharif season produces the main harvest, provided the summer monsoon brings the proper rainfall.

In recent years, the rains have failed frequently and harvests have been smaller. Crops also suffer in the eastern part of the country when the Ganges and its branches flood the plains.

Recent developments

India is a land of small farms that use family labor and work animals to till the land. In recent years, the government has introduced new varieties of seeds, as well as new ways of using fertilizer and modern irrigation systems. These changes have helped Indian farmers increase their crop production.

Tea plants grow throughout India, *left,* but the best-producing areas are in southern India and Assam. The well-known Darjeeling tea grows on hillsides near the city of the same name. India produces 1-1/2 billion pounds (670 million kilograms) of tea a year.

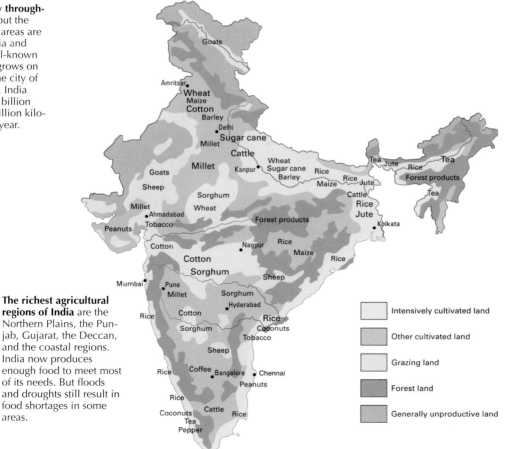

The richest agricultural regions of India are the Northern Plains, the Punjab, Gujarat, the Deccan, and the coastal regions. India now produces enough food to meet most of its needs. But floods and droughts still result in food shortages in some areas.

Intensively cultivated land

Other cultivated land

Grazing land

Forest land

Generally unproductive land

Economy

India has a large economy, which is determined by its *gross national product* (GNP), the value of all goods and services produced in a year. But when the country's huge population is divided into its GNP, the GNP *per capita* (per person) is very low. For this reason, India is considered a developing country.

Since India became independent, the government has worked very hard to stimulate industrial growth. Today, industrial production is six times as great as it was in 1950.

Although India is now one of the largest industrial nations in the world, the country still faces many challenges in its economic development. Most of its rich natural resources—farmland and mineral deposits—have not been sufficiently developed. And because the huge population keeps increasing so rapidly, it is difficult for economic gains to make a real difference.

Manufacturing

The clothing and textile industries employ more Indian workers than any other industry. Cotton mills are located in the cities of Mumbai and Ahmadabad, Punjab has woolen mills, and Kolkata has jute factories.

Since 1950, the Indian government has built major iron and steel mills at Bhilai, Bokaro, Durgapur, and Raurkela. These mills were built with aid from the Soviet Union, the United Kingdom, and West Germany and are controlled by the Indian government.

Indian factories use iron and steel to make automobiles, bicycles, diesel engines, appliances, military equipment, and many kinds of industrial machinery. The assembly of electronic products is a growing industry in India today.

In December 1994, more than 2,000 people died when poisonous gas leaked from a pesticide factory at Bhopal, in central India. The factory was owned by an American company, Union Carbide Corporation. About 200,000 people were injured by the leak.

To continue its industrial growth, India must develop its natural resources further. Industry needs energy to develop but India is at a disadvantage because it must import large quantities of oil. Much of the country, particularly the rural areas, has no electricity supply.

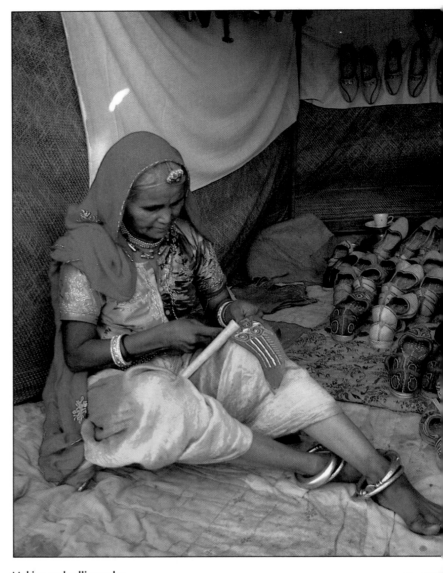

Making and selling colorful handicrafts provides a living for many Indians. People all over the world buy the woven cloth, woodcarvings, embroidered shawls, carpets, and other items made by Indian weavers and craft workers.

India's motion-picture industry is the world's largest. Movies provide entertainment for many Indian people because few families own television sets. Dramatic versions of Hindu epics and romantic musicals are the most popular films.

India's busy textile industry produces clothing made of home-grown cotton and jute, as well as wool, rayon, and colorful silks. Clothing and textile manufacturing employs more Indian workers than any other industry.

Nuclear power plants provide some of India's electricity. In 1963, the United States supplied the first such plant, near Mumbai.

Rich natural resources

India's natural resources include a variety of raw materials suitable for industrial development. India produces about 5 per cent of the world's iron ore. The country also has a good supply of coal, especially in the states of Bihar and West Bengal. Petroleum is produced by offshore rigs near Mumbai and oil wells in Assam.

Deposits of mica are also important to India's economy. Mica is used in the manufacture of electrical devices, and India provides much of the world's supply.

India's other mineral resources include large deposits of bauxite, beryllium, chromite, gypsum, limestone, magnesite, natural gas, salt, and titanium.

Cottage industries

Millions of Indian people throughout the country work at home, employed in what are called *cottage industries*. These workers make many beautiful items.

Home weavers create fine fabrics of cotton, rayon, and silk by hand, as well as beautifully designed carpets and rugs. They spin fine laces of gold and silver threads. Home craft workers also make brassware, jewelry, leather goods, and woodcarvings.

Looking to the future

India has made much progress in its economic development in the years since independence. Today, India has more scientists and skilled workers than most of the other countries in the world.

To continue its modernization programs, India needs money. In an effort to boost the economy, the government now encourages other countries and private companies to invest in India's industry.

People

For centuries, many different groups of people migrated to India from other parts of Asia. Today, the descendants of these ancient peoples give the population of India its great variety of ethnic and language groups.

Most people of northern India are descended from the Aryans, an ancient people who invaded the area about 1500 B.C. Descendants of the darker-skinned Dravidians live in the southern part of the country. Dravidians were among the earliest known inhabitants of India.

Muslim people also settled in India. Their descendants are concentrated in the northeast states of Bihar, Uttar Pradesh, and West Bengal. Some groups who live in the far north and northeast are closely related to peoples of East and Southeast Asia.

India is also home to more than 70 million tribal people who live in remote forests and hills. The largest group, the Gonds, live near the Eastern Ghats. Other tribal groups include the Bhils, Khasis, Nagas, Oraons, and Santals.

Many languages

Sixteen major languages and more than 1,000 minor languages and dialects are spoken in India. The major languages fall into two groups—the Indo-European and the Dravidian.

About 73 per cent of the people speak the Indo-European languages, which include Hindi—India's most widely spoken language. Also included in this group are Urdu, which is closely related to Hindi, Assamese, Bengali, Gujarati, Kashmiri, Marathi, Oriya, Punjabi, and Rajasthani. These languages come from Sanskrit, the ancient Indian language.

The Dravidian languages, spoken mainly in the southern part of India, include Kannada, Malayalam, Tamil, and Telugu.

Two other language groups found in India are the Sino-Tibetan languages spoken in the northern Himalayan region and near the Myanmar border, and the Mon-Khmer languages spoken by some ethnic groups in eastern India.

The principal official language of India is Hindi. Sanskrit and 13 regional languages are also official languages while English is considered an "associate" national language. At least one major regional language is spoken in each state. Elementary and high school students study in their regional language and learn Hindi as a second language. In most colleges and universities, classes are taught in regional languages, but English is widely used.

Because so many languages are spoken in India, communication and understanding between ethnic groups has often been difficult. However, a more serious problem facing the Indian people today is overpopulation.

The population problem

India is the second most-populated country in the world. In recent years, due to improved sanitation and health care, the death rate has dropped more rapidly than the birth rate. During the 1970's and 1980's, India's population increased by up to 14 million people a year.

Overpopulation has caused serious overcrowding throughout India, particularly in the cities. In Kolkata, for example, the population density averages about 110,000 people per square mile (42,500 per square kilometer).

Many Indian men wear turbans of various shapes and colors, especially during celebrations. These men, *right*, have been taking part in the Hindu festival Holi by throwing colored powder and water on one another.

A Muslim *Sufi,* *above,* wears the traditional woolen garments for which this Islamic sect was named. *Sufi* comes from the Arabic word for wool.

The Toda people, *below,* live in small, thatched houses nestled in the Nilgiri Hills of the southern state of Tamil Nadu. They herd buffalo and make cane and bamboo articles. Very few Toda survive today.

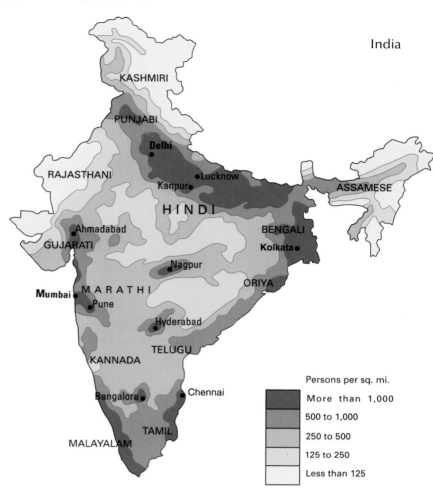

Persons per sq. mi.

More than 1,000

500 to 1,000

250 to 500

125 to 250

Less than 125

Schoolchildren in the southern seaport of Chennai often eat a lunch of traditional vegetarian food. A typical meal consists of grains and spiced vegetables. Most Hindus do not eat beef, and some eat no meat at all.

Millions of people live in slum dwellings. Often, an entire family inhabits a one-room shelter made of scraps of wood or metal. Many others are so poor that they have no home at all and must sleep in the streets.

About 40 per cent of India's population is made up of children aged 14 or under, and schools are overcrowded. Many children must drop out of school to help support their families. Only about a third of India's adults can read and write. Their lack of education limits their job opportunities and, as a result, they continue to live in poverty.

To help control population growth, the government has introduced programs that encourage people to have smaller families.

This young Rajasthani, *left,* is a member of the Aryan ethnic group. Most Rajasthanis speak an Indo-European language that comes from Sanskrit.

623

Culture

About 80 per cent of India's people practice the Hindu religion, one of the world's oldest living religions. Hindu rules and customs have an important influence on Indian life.

Unlike many other religions, Hinduism was not founded on the teachings of one person. Many different cultural, racial, and religious groups had a role in shaping Hindu philosophy.

The caste system

A basic part of Hinduism is the *caste* system. Caste is a strict system of social classes. The ancient Aryans, who invaded India about 1500 B.C., used the caste system to limit their contact with Dravidian people who had already settled there.

The Hindu castes are grouped into four main categories, called *varnas.* In order of rank, these varnas are (1) *Brahmans,* the priests and scholars; (2) *Kshatriyas,* the rulers and warriors; (3) *Vaisyas,* the merchants and professionals; and (4) *Sudras,* the laborers and servants.

About 20 per cent of the Indian population are ranked below the lowest Sudra caste. These people are called *untouchables,* and they have traditionally held the lowest jobs.

According to Hindu belief, membership in a caste is established at birth and is difficult, if not impossible, to change. A person's social status in the community depends on the caste to which he or she belongs. Each caste also has a traditional occupation.

There are thousands of castes in the caste system, each with its own rules of behavior. Friendships and marriages rarely occur between members of different castes.

Some people believe that the caste system slows India's progress toward becoming a modern nation. In recent years, however, many caste barriers have broken down. People of different castes work in the same offices and factories and mingle in public places. India's 1950 Constitution gave untouchables equal rights as full citizens.

Even with these changes, few Indians want the caste system to die out completely. They prefer the security of knowing exactly where they belong in society. Also, caste organizations preserve traditional skills from generation to generation and provide help to needy members.

The Buddhist wallpainting from which this detail comes was painted in the cave temples of Ajanta, Maharashtra, during the 100's B.C. Although Buddhism was once the chief religion of India, less than 1 per cent of Indians are Buddhists today.

Amritsar's Golden Temple, *above,* set in a sacred pool, is the holiest Sikh shrine. Sikhism was founded about 1500. About 14 million Sikhs live in India.

Hindu worship

Hindu belief and conduct is based on the teachings of Sanskrit literature. *The Vedas* are the oldest Hindu scriptures. These philosophical works existed for centuries and were passed down orally from generation to generation before they were written down.

Many ancient Hindu rituals are still widely observed. Millions of Hindus visit temples along the Ganges River, the most sacred river in India. Hindu temples hold annual festivals to honor events in the lives of their gods and most Hindus have a shrine in their home devoted to a particular god.

Hindus believe that animals as well as human beings have souls. They have reverence for cows, monkeys, and other animals. Although not all Hindus are vegetarians, most will not eat beef.

Ganesh, the elephant-headed Hindu god, is featured in this Tamil Nadu temple shrine. Ganesh is universally honored by Hindus as the "remover of obstacles." Hindus worship many *divinities* (gods and goddesses).

Hindu reverence for the cow dates from about 3,500 years ago. The sacredness of the cow is revealed in Hindu scriptures, particularly in the stories of Krishna among the *gopis* (milkmaids) of Brindaban. Krishna is a Hindu god. His conversations with the Pandava warrior Arjuna discuss the meaning and nature of existence in the philosophical work *Bhagavad-Gita*. Reverence for cows is also part of the Hindu philosophy of *ahimsa*, noninjury to living creatures.

Minority faiths

Most Indians are Hindus. However, Muslims, Christians, and other religious groups also live in India.

About 14 per cent of the Indian population are Muslims, making Islam the second largest religion in India. Most Muslims live in the northern part of the country. Christians make up about 2 per cent of the population. Many live in the state of Kerala and in the areas along India's northeast border.

Sikhs, whose religion is a combination of Hinduism and Islam, make up about 2 per cent of the population. They are mainly wheat farmers in the north. Buddhists and Jains each make up less than 1 per cent of the population.

The Hindu Dussehra festival, celebrated here in Delhi, commemorates events in the lives of the divinities. These festivals attract huge crowds of Hindus, who come to worship, to pray, and to enjoy the colorful display.

The River Ganges

The Ganges is the most important river in India and one of the largest in the world. It begins as a pool in a Himalayan ice cave 10,300 feet (3,139 meters) above sea level.

The river flows southeast through the northern part of India and into Bangladesh. There, some of the branches of the Ganges join the Brahmaputra River to form a large delta and together they flow into the Bay of Bengal.

The Ganges is one of India's greatest natural resources. As the river flows through the country, it leaves soil deposits that make the land rich and fertile for farming. Through irrigation systems, its waters provide moisture for India's crops.

In the past, when the Ganges River carried many merchant ships, some of India's largest cities were built along its banks. In recent years, irrigation has drained much of its water and steamboats can navigate only the lower river. As a result, the river has become less important for trade.

To the people of India, however, the Ganges—or Ganga, as the Hindus call it—has far more importance for its spiritual value. Hindus consider it the most sacred river in India. They believe that bathing in the Ganges just once is a greater assurance of salvation than a lifetime of prayer and meditation.

The goddess Ganga

Ancient Hindu scriptures tell how, for a very long time, the goddess Ganga flowed only in the heavens. She was an object of great beauty and sacredness but she was of little use to the dry earth below.

The Maharaja Sagar, an Indian king of great riches, had 1,000 sons who traveled far and wide in search of new lands to conquer. During their journey, they came upon a Hindu wise man named Kapil. Annoyed because the sons had disturbed his meditation, Kapil burned them to death with his eyes.

When the maharaja heard what had happened to his sons, he begged Kapil to return them to life. Kapil agreed to restore the sons' lives, but the goddess Ganga would have to come down from the heavens and touch their ashes.

From its beginning in northern India, the Ganges flows southeast through India and Bangladesh for 1,540 miles (2,478 kilometers) and empties into the Bay of Bengal. The Ganges has several *tributaries* (branches), including the Jumna, Ramganga, Gomati, Ghaghra, Son, and Sapt Kosi.

Ganga was afraid that her mighty torrent of water would shatter the earth's foundation. She asked the god Shiva to stand above the earth on the rock and ice of the Himalaya.

Ganga then flowed down to earth through Shiva's matted hair, which absorbed the shock of her waves. She trickled down from the mountains and across the plains, bringing water and life to the dry earth. Ganga then touched the ashes lying there, and the maharaja's sons came back to life. Thus, the earthly Ganga was born.

Ritual bathing

The holy cities on the banks of the Ganges attract thousands of Hindu pilgrims. They bathe in the river and take home some of its water. The river is said to bring purity, wealth, and fertility to those who bathe in it. The Hindus have built *ghats* (stairways) along the banks of the Ganges in order to reach the water more easily.

Some pilgrims bathe in the Ganges to cleanse and purify themselves. The sick and crippled bathe in the Ganges in the hope that the sacred waters will cure their illnesses. Others come to die in the river, because the Hindus believe that those who do will be taken to Paradise.

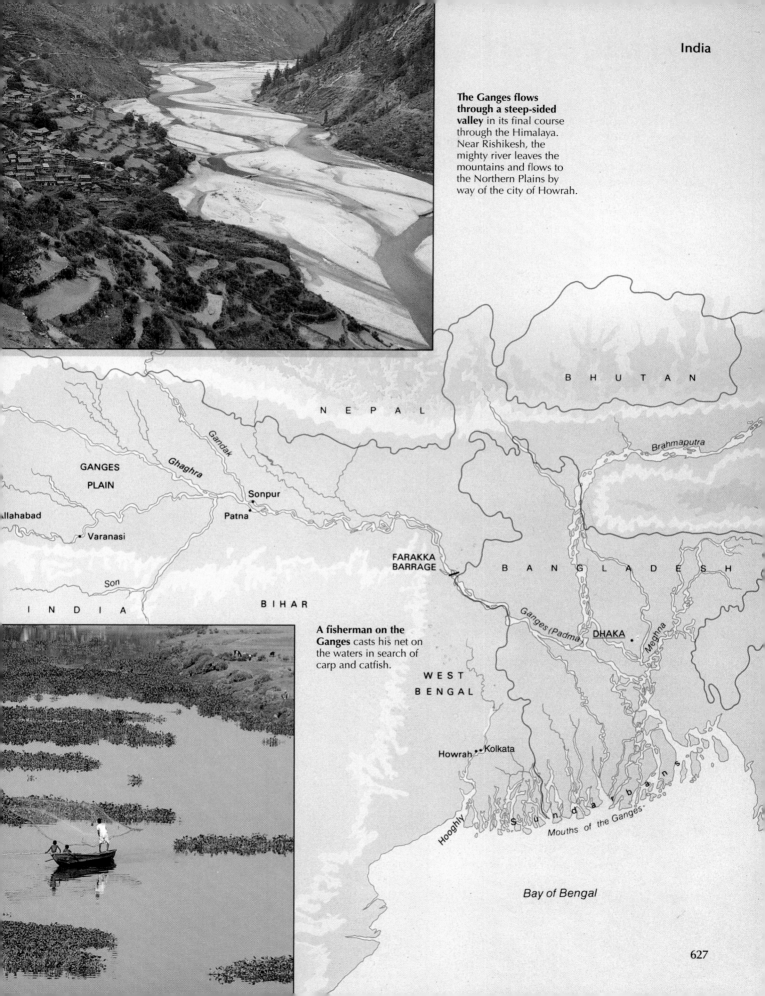

The Ganges flows through a steep-sided valley in its final course through the Himalaya. Near Rishikesh, the mighty river leaves the mountains and flows to the Northern Plains by way of the city of Howrah.

B H U T A N

N E P A L

Brahmaputra

GANGES
PLAIN

Gandak

Ghaghra

Allahabad

Sonpur

Patna

Varanasi

B A N G L A D E S H

Son

FARAKKA
BARRAGE

Ganges (Padma)

DHAKA

Meghna

I N D I A

B I H A R

A fisherman on the Ganges casts his net on the waters in search of carp and catfish.

WEST
BENGAL

Howrah • Kolkata

Hooghly

S u n d a r b a n s

Mouths of the Ganges

Bay of Bengal

627

Cities and Villages

India has been described as a land of villages. About 72 per cent of its people live in about 557,000 villages. Most of these villages have less than 1,000 people. By contrast, India has about 4,000 cities and towns. Only about 225 cities have populations over 100,000.

Since the 1980's, the population of the cities has grown dramatically. Many villagers leave the rural areas to look for work in cities, where wages are higher.

As a result, India's urban population is growing at about twice the rate of its rural population. Urban growth is concentrated in Kolkata, Delhi, Chennai, and Mumbai.

Village life

Indian villages, like the people who live in them, look quite different from one part of the country to another. The stone dwellings of the chilly mountain regions are quite unlike the bamboo and matting huts of the hot central and southern areas. Some are small settlements around isolated farms, while others are tightly knit communities built around larger farms.

Life is simple for Indian villagers. Most live in clusters of huts made of mud and straw. These huts usually have mud floors and only one or two rooms. Household articles may include brass pots for cooking, clay pots for carrying water and storing food, and little else. About half of India's villages have access to electricity, although individual homes may not have electric power. Many people use kerosene lanterns.

Many village dwellings have no running water. The women get water from the village well, often the center of activity. They pour the water into pots and carry the pots home on their heads.

Meals usually consist of rice and *dal,* a porridge made of *pulses,* which are seeds of such pod vegetables as beans, chickpeas, pigeon peas, and lentils.

Family ties are important to the villagers. Marriages are thought of as a relationship between two families, rather than a union between two individuals. An Indian household may include not just parents and children, but also the sons' wives and their children. Relatives and neighbors often join

In Chandni Chowk, *left,* or Silver Square, the commercial center of Old Delhi, merchants sell silver jewelry, wholesale goods, sweetmeats, handicrafts, and clothing in market stalls.

An artificial reservoir, *right,* known as the *village tank,* is often the heart of thousands of small communities throughout India. Many were built centuries ago and may be a village's only source of water.

In the slums of Mumbai, *below left,* many people live in shacks made of wood or metal scraps. Others are crowded into high-rise tenement buildings. The slum areas have poor water supplies and sanitation.

together to help those who have met with some misfortune.

Rural life has improved in many Indian states. In Haryana and Punjab, for example, economic growth has brought such modern conveniences as electricity, improved drinking water and sanitation, and paved roads.

City life

The cities founded by the Europeans—Kolkata, Chennai, and Mumbai—have become major urban centers. These cities usually have two separate areas—a British section and an Indian section.

The British section has modern buildings, shopping districts, and wide, treelined streets. Most of the wealthier Indians, and those with high military rank, live in this section. Many of these Indians have adopted Western ways and live in comfortable bungalows.

The Indian section is quite different. The narrow, twisting streets are crowded with bicycles, carts, animals, and people. Millions of people live in terrible slums, where water supplies and sanitation are poor.

Village children attend school at Naricanda in the Himalayan foothills. Although government education programs have improved the country's literacy rate, only a third of India's adults can read and write.

The Holy City of Varanasi

Along a sandy ridge on the west bank of the Ganges River lies the city of Varanasi, also known as Banaras or Benares. One of the largest cities in the northern Indian state of Uttar Pradesh, Varanasi is known for the fine silk fabrics made by its craft workers. For India's Hindus, however, Varanasi is a place of deep religious importance.

Holy bathing

The Hindus believe that the whole length of the Ganges is sacred, and that Varanasi is the holiest place on earth. Each year, a million Hindu pilgrims from all over India come to the city to bathe in the waters. Many also come to Varanasi to die, because Hindus believe that eternal salvation comes to those who die in the Ganges.

Pilgrims enter the Ganges by way of one of Varanasi's 52 *ghats* (steps leading down to the river). The ghats extend about 5 miles (8 kilometers) along the riverbanks. Five of these ghats are reserved for holy bathing. Each day, before the light of dawn, devout Hindus begin to gather for their holy bath.

Three ghats are reserved for cremating the dead. More than 3,000 cremations take place every year in Varanasi at the ghats of Jalasayi, Harish Chandra, and Manikarnika.

The oldest of the ghats, Manikarnika takes its name from the *manikarnika* (earring jewel), which belonged to Sati, wife of the Hindu god Shiva. According to Hindu scriptures, Brahman priests of the highest caste found Sati's lost earring but did not return it. Shiva punished the Brahmans by lowering their caste status to those of *doams*. He put the doams in charge of the burning ghats. Today, the doams still tend the funeral pyres at the burning ghats.

Religious and cultural center

Varanasi is sacred to the Buddhists as well as to the Hindus. It was in Sarnath, just a few miles from Varanasi, that the Buddha preached his first sermon to five holy men. During his sermon, the Buddha preached the message of how to overcome suffering. The delivery of this sermon is one of Buddhism's

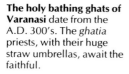

The holy bathing ghats of Varanasi date from the A.D. 300's. The *ghatia* priests, with their huge straw umbrellas, await the faithful.

Hindus believe that bathing in the Ganges, *above,* cleanses the soul. Pilgrims come to Varanasi from all over India to perform this ritual.

most sacred events, and Varanasi is one of Buddhism's most sacred places.

The city is also a center of culture and learning. Banaras Hindu University and other colleges are located there. Varanasi is also the site of a library containing more than 150,000 ancient manuscripts and other documents.

Scholars and sages, from Shankaracharya (700?-750?) to Vivekananda (1862-1902), have found inspiration in Varanasi. The city is also home to Ravi Shankar, the world's best-known sitarist. A *sitar* is a stringed instrument used to play the classical music of India.

The arts and crafts of Varanasi have been prized since ancient times. In addition to their famous silk fabrics, the city's craft workers make shawls, *saris* (long dresses worn by Indian women), gold-embroidered cloth, hand-hammered brassware, and heavy gold and silver jewelry.

Hindu pilgrims flock to the city of Varanasi to purify themselves in the waters of the Ganges. Devout Hindus expect to come to Varanasi at least once in their lives.

A funeral pyre goes up in flames at a burning ghat while another body is prepared for cremation. The task of cremation is given to the *doams,* a Hindu caste.

Exploring India

India has fascinated Westerners for centuries. To early European explorers, traders, and adventurers, India was a land of mystery and excitement. Today, visitors speak of its timeless magic, where past meets present at every turn.

It is also a land of great variety in people, cultures, and landscapes. In no other country would a visitor be greeted by a hotel clerk wearing a *dhoti* (a simple white garment wrapped between the legs) and displaying a caste mark on his forehead. India is also the only country in the world where a traffic jam could be caused by a sacred cow deciding to sit down in the middle of a city street.

Indian leaders understand the value of tourism to India's economy. In recent years, they have improved tourist facilities and opened up new areas of the country to visitors.

Because India is so vast, a typical tour of 10 days to 3 weeks is not long enough to see even a large part of the country. As a result, most tours take in only a single region. Even so, visitors have a variety of places from which to choose. They may enjoy the glamour and bustle of city life in Mumbai or Kolkata, the sun-drenched beaches of Goa, or the peacefulness of the lake regions of Kashmir.

Northern India is most enjoyable in the cool, mild months of October through February. In southern India, where it is almost always hot, November through January is the coolest time.

Wherever tourists go in India, they are welcomed by the generous nature and friendliness of the Indian people. Visitors are advised to return the traditional Indian greeting of *namaste*—placing palms together in front of the chest and bowing. The Indian people prefer not to shake hands.

For visitors to northern India, the city of Delhi is a good place to start. Delhi is India's second largest city and features many historic landmarks.

The Red Fort, one of the city's most impressive monuments, is a red sandstone structure that covers several blocks. It was built in 1648 by the Mogul Emperor Shah Jahan. The remains of the imperial palace and other Mogul structures lie within its walls.

India's railroad system is the most inexpensive and convenient way to see the country. In Kurseong, West Bengal, the train runs so close to the road and shops and moves so slowly that merchants trade with passengers along the way.

The state of Tamil Nadu, *far right,* in southern India, has many Hindu temples. Hindu sculpture dating from the 600's may be seen at the Seven Pagodas in Mahabilipuram, south of the seaport city of Chennai.

Carved from very hard sandstone, the detailed latticework of the Maharaja's Palace in Jodhpur, *bottom right,* shows the great skill and patience of the ancient stonemasons.

The lake palace in Amber, *bottom,* just outside Jaipur in the northwestern state of Rajasthan, was built by Mogul rulers in the 1600's. Its peaceful setting on a steep hill skirted by a lake shows the beauty of rural India.

While the Mogul Empire left its mark on Delhi, the British influence can be seen in New Delhi. Now the capital of India, New Delhi was built in the early 1900's when India was part of the British Empire. It features wide, treelined avenues and many gardens.

South of Delhi lies the city of Agra, the site of India's most famous monument—the Taj Mahal. One of the most beautiful and costly tombs in the world, the Taj Mahal was built by Shah Jahan. Construction began about 1630 and took more than 20 years to complete.

The island city of Mumbai, formerly called Bombay, is situated on the west coast of India. In its harbor stands Elephanta Island, with its cave temples dating from the 700's. East of Mumbai, near Aurangabad, the Buddhist monasteries and shrines in the caves of Ajanta are decorated with frescoes (wall-paintings) from the Gupta period.

South of Chennai on India's east coast stands the sacred Hindu city of Kanchipuram. A capital of the Dravidian dynasties from the 700's to the 1300's, Kanchipuram is the site of some 150 Hindu temples.

Project Tiger

At the beginning of the 1900's, about 40,000 tigers roamed the jungles and grasslands of India. In 1972, the Indian government's first formal tiger census found only about 2,000 tigers left in the entire country. Once common throughout southern Asia, the tiger had become an endangered species.

The low population of the tigers alarmed conservationists and wildlife experts. The year after the tiger census, Prime Minister Indira Gandhi launched "Project Tiger," thought by many to be the most dramatic rescue operation in conservation history.

Project Tiger began on April 1, 1973, at the Corbett National Park in the northern state of Uttar Pradesh. It was the first of nine original tiger preserves created through the program.

These large preserves provide enough space for a natural tiger habitat, while also protecting the human population who live close by. Corbett National Park covers 200 square miles (520 square kilometers).

By the mid-1990's, 23 tiger reserves had been established.

Project Tiger has been very successful. According to official estimates, the tiger population doubled in the program's first 10 years of operation.

Tiger numbers began to decline again, however, after the program's early successes. Poachers began snaring and poisoning tigers. It was reported that tigers were being killed so that their bones could be used in medicines. By the early 1990's, it had become increasingly difficult for tourists and trackers to spot tigers, and conservationists renewed their fund-raising and publicity efforts.

Threatened with extinction

Several factors have contributed to the drop in India's tiger population. Farmers have cut down the forests in which many tigers live to make room for grazing land and to obtain wood. The tiger is now competing for living space with millions of Indian people.

Hunting has also helped to make the tiger an endangered species. In the 1950's and 1960's alone, hunters in search of trophies and skins killed more than 3,000 tigers. A government ban on shooting tigers went into effect in 1970. In 1975, the Convention on International Trade in Endangered Species (CITES) passed an international law making it illegal to trade in the skins of endangered species.

Creating the reserves

Because tigers are large predators, they need a great deal of land for their habitats. Therefore, the leaders of Project Tiger created large reserves that give tigers enough room to live, hunt for prey, and raise their cubs.

Each reserve consists of a *core area,* which people are forbidden to enter and where nature is allowed to take its own course. Surrounding the core area is a *buffer zone,* which people are allowed to enter for a specific reason, such as to cut bamboo. The buffer zone makes up about 60 per cent of the reserve.

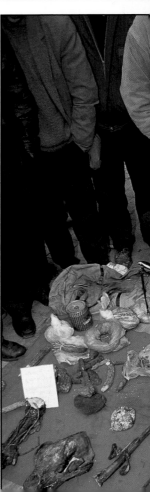

Saving many animals

The success of Project Tiger is important for many reasons. The tiger has been saved from extinction—at least for now. Other species also benefit by the preservation of the tiger's large natural habitat.

India has a wealth of magnificent wild creatures. Many of these, including the elephant, the Indian rhinoceros, the gaur, and the barasingha, are rare or endangered species. Some of these species now live in Project Tiger reserves with no interference from human activity.

Project Tiger has also taught people around the world that humankind and nature can—and must—live in harmony.

A tiger cools off in the water, *left.* Tigers often soak themselves for an hour or so in marshes and rivers. Tigers are good swimmers and may swim across rivers in search of prey.

A monkey becomes a meal for a tiger in the Corbett National Park. Although they prefer large prey, tigers will also feed on such small prey as peafowl, monkeys, tortoises, frogs, and porcupines.

The illegal sale of tiger parts, *left,* continues, despite increasing awareness of the tiger's plight.

Scientists attach a monitoring device to a tranquilized female tiger. These devices are used to track the movements of animals in Project Tiger reserves, allowing scientists to learn more about their habits.

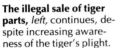

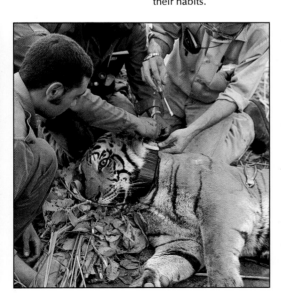

Sikkim

Tiny Sikkim lies on India's northern border with China. It is nestled between the Himalayan kingdoms of Nepal in the west and Bhutan in the east.

Sikkim is the second smallest Indian state. It covers an area of 2,740 square miles (7,096 square kilometers) tucked away high in the Himalaya. Nepal lies to the west of Sikkim, China to the north, Bhutan to the east, and Bangladesh to the south.

Before Sikkim became a state of India in 1975, it was a monarchy ruled by a *chogyal* (king). Sikkim first became an independent monarchy about 1640, when Penchu Namgyal was crowned chogyal. At that time, Sikkim controlled lands that are now part of Bhutan, China, India, and Nepal.

In 1780, warriors from Nepal and Bhutan invaded Sikkim and seized much of the land. The British returned some of this land to Sikkim when they conquered the Nepalese in 1814.

Sikkim came under the protection of the British in 1861. Later, a British official took over much of the chogyal's power. By 1918, the chogyal had regained control over internal matters.

The British gave control of Sikkim to India when India became an independent nation in 1947. The Indo-Sikkim Treaty of 1950 gave India much of the same powers that Great Britain had before 1947. Sikkim was independent in its internal affairs, but India controlled its foreign relations, defense, and communications systems.

In April 1975, a special referendum was held in Sikkim, and voters approved the legislature's proposal to become a state of India. In May, Sikkim officially became an Indian state.

Landscape and economy

Although Sikkim is relatively small in area, its landscape is extremely varied. Towering mountains extend in an arc along its western, northern, and eastern borders. Mount Kanchenjunga, the third highest mountain in the world, lies on the western border with Nepal.

Grassy meadows surround Gangtok, Sikkim's capital and its only city. Gangtok is located in the southeast part of the state along the route used for centuries by major caravans between Tibet and India.

Religious processions are a common sight in Sikkim. The Lepchas and Bhutias practice Lamaism. The Nepalese of Sikkim are Hindus.

Lamaist monks, *below*, often enter the order when they are very young. They are devoted to the teachings of Buddha, the founder of Buddhism who lived in the 500's B.C.

The southern border opens to a broad valley watered by the Tista River, the state's major waterway. In the river valleys, thick tropical rain forests cover the land.

Farming is the basis of the Sikkim economy. Crops include rice, corn, and other cereals. Apples, cardamom, citrus fruits, pineapples, and potatoes are also grown.

Handicrafts are the chief Sikkim industry. Skilled craft workers weave cloth, blankets, and rugs. Sikkim is also known for its copperware and woodcarvings.

Way of life

Most of the people who live in Sikkim are Nepalese, Lepchas, and Bhutias. The Nepalese, who make up about 70 per cent of the population, live mainly on small farms in the southern regions. They work the land with simple hand tools. Most Nepalese speak Nepali. They are Hindus, but their religion is strongly influenced by Buddhism.

The Lepchas, the first settlers in Sikkim, now live in distant valleys, hunting and fishing for food. Some Lepchas also farm and raise livestock. Most speak Sikkimese.

The Bhutias are a seminomadic people. In summer, they live in tents and herd cattle and *yaks* (Asian oxen) in the high mountain meadows. In winter, they live in wooden houses in the highlands. Like the Lepchas, the Bhutias speak Sikkimese. Both groups practice Lamaism, also known as Tibetan Buddhism.

Goa

GOA

Goa is a former Portuguese colony on the west coast of India, now an Indian state. About 1,002,535 people live in Goa, which is known for its fine beaches and warm, tropical climate. Its capital city is Panaji.

The beach at Baga, Goa, *right,* is a beautiful backdrop for an outrigger fishing boat at sunset. Fresh fish are a favorite dish in many local hotels and restaurants.

The highly decorated church of Saint Cajetan was modeled after St. Peter's Basilica in Rome. The church's *crypt* (underground chamber) holds the tombs of generations of Portuguese rulers. The church features a large arched window with tiny panes made of sea shells.

Rural Goan families, *center,* suffered great hardship under Portuguese rule. Conditions have improved since India took control of Goa in 1961. Today, government programs provide better housing and more effective health care.

Many young travelers, such as this backpacker, come to Goa for its relaxed atmosphere and friendly inhabitants. Because Goa's economy depends on tourism, many tourist facilities are available.

In a small corner of the west coast of India, the clear blue waves of the Arabian Sea roll slowly over the white sands of a sun-drenched beach. Warm sea breezes blow inland across the rooftops of an old fishing village. This is Goa, a tiny paradise of sandy beaches, sleepy villages, and warm, tropical weather.

Now an Indian state, Goa was formerly a Portuguese colony and later a territory of India. It covers an area of about 1,430 square miles (3,702 square kilometers) south of Mumbai between the Western Ghats and the Arabian Sea.

Early history

Goa's original inhabitants were a Dravidian people called the Kannadigas. The ancient Hindu city of Goa was famous for its beauty, wealth, and culture. References to Goa appear in the early Hindu epic work, the *Mahabharata.*

During the Middle Ages, European traders sought a foothold on the west coast of India. They wanted to use the region as a source of exotic spices and as an important link on the Arabian trade routes.

The Portuguese were the first Europeans to appear on the shores of Goa. Alfonso de Albuquerque arrived in 1510. He conquered the Indians and soon controlled the coast. The region became a powerful and wealthy trading area, where silks, spices, porcelains,

and pearls passed in and out of Goa's harbors. The city became known as *Golden Goa* throughout the civilized world.

Thirty-two years after Albuquerque arrived in Goa, Saint Francis Xavier came to Goa as a Christian missionary. Although Saint Francis Xavier left Goa to preach in distant lands, his body was returned after his death in 1552. He is buried in the *Bom Jesus Basilica* in the city of Old Goa. Today, about 40 per cent of the Goan people practice an easternized version of Roman Catholicism.

Goa reached the height of its power in the 1500's. By 1750, the capital of Old Goa had been nearly destroyed in a series of invasions by the Hindus, Muslims, Dutch, and British. Portuguese rule continued into the early 1900's, as Goa became a center for illegal trading activities.

When India became independent in 1947, the government tried to persuade the Por-

tuguese to withdraw from Goa. In 1961, the Indian army invaded Goa and ended 451 years of colonial rule. Goa was first declared a Union Territory along with the tiny colonies of Daman and Diu. In 1987, Goa became a separate state of India.

Goa today

The strongest element of Goa's economy today is international tourism. Visitors from all over the world come to enjoy its sunny climate and beautiful beaches. Goa is also a major industrial center because of its rich deposits of manganese and other mineral ores.

Goa has lost much of its male population to Africa, the Middle East, and the large cities of India. Those men who have left did so largely to find employment. Older people, women, and children have been left behind to tend the small farms, which are little more than small patches of land.

Indian Islands

In addition to its huge peninsula, the country of India also includes two island territories. These are (1) the Union Territory of Lakshadweep, which includes the Amindivi Islands, Minicoy Island, and the Laccadive Islands; and (2) the Union Territory of the Andaman and Nicobar Islands.

The islands of Lakshadweep rise in the Arabian Sea 124 miles (200 kilometers) off the southwest coast of India. The Andaman and Nicobar Islands are situated in the Bay of Bengal, 744 miles (1,200 kilometers) off the southeast coast of India.

Lakshadweep

Even though the name *Lakshadweep* means *thousand islands,* there are only 27 islands in the chain. Their total area is 12 square miles (32 square kilometers). Most of the 40,200 people in the territory live on the Laccadive Islands. The rest live on the Amindivi Islands.

Most of the islanders belong to various Arabian tribes. They speak the Malayalam language, a Dravidian language commonly used on India's southwest coast.

The Union Territory of Lakshadweep consists of the Amindivi and Laccadive Islands, and Minicoy Island.

Fishing boats and nets form a pattern on the skyline at Cochin on India's mainland. Cochin is the main port of call for traders from the Lakshadweep Territory. The islanders bring coconut fiber and mats to trade for rice.

This engraving from the 1800's shows a paddle steamer approaching the Andaman Islands while islanders attempt to defend their land from invasion. Port Blair on South Andaman was a British penal colony from 1858 to 1945.

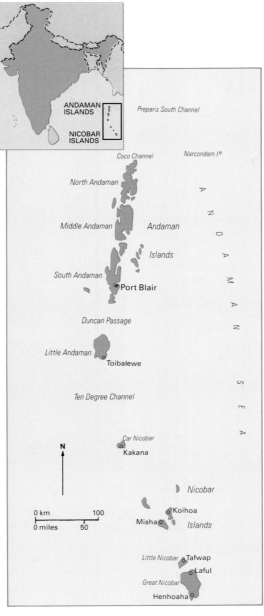

The Andaman and Nicobar Islands enjoy the protection of the Indian government as a Union Territory of India.

Very few descendants of the original Andaman Islands inhabitants remain today. The survivors live by hunting and fishing. They also collect honey from the wild bees in the islands' dense tropical rain forests.

The islanders live a simple life. The women make an elastic fiber called *coir* from coconut husks. Coir is used in the manufacture of matting. The men build boats and trade coir, processed fish, palm oil, and vegetables for rice. Rice is their staple food.

In the past, the islands of Lakshadweep were India's western outpost and served as a navigation point between Africa, Arabia, and India. Today, the islands are slowly opening up to tourism.

The Andaman and Nicobar Islands

Like Lakshadweep, the Andamans and the Nicobars have the status of Union Territory. They have been called *Marigold Sun* for their natural beauty. The 223 islands dot an area in the Bay of Bengal covering 3,185 square miles (8,249 square kilometers).

Of these 223 islands, 204 belong to the Andamans, including the "big" islands of North, Middle, South, and Little Andaman. The largest Nicobar Islands are Greater and Lesser Nicobar, Kamorta, Katchall, and Kar Nicobar. Port Blair, a city of about 55,000 people on South Andaman, is the territory's seat of government and commercial center.

The landscape of the islands is marked by high mountain peaks and dense rain forests. Only the narrow coastal plains, valleys, and the mouths of rivers are suitable for settlement. Because of the great demand for land on the islands, forests are being cleared at a rapid rate.

The people of the Andaman and Nicobar Islands belong to a variety of ethnic groups. The Indian islanders are the descendants of convicts who arrived in the 1800's when Port Blair was a British penal colony. Port Blair was originally intended to hold the thousands of Indians seized during the Sepoy Rebellion of 1857-1859.

Migrants and refugees from East Pakistan (now Bangladesh) also settled on the islands. The arrival of these outsiders greatly reduced the numbers of original inhabitants, who are of Negrito ancestry. These people are now nearly extinct. The survivors live in forest reservations on the main islands, or in voluntary isolation on the islets and in the mountains. The islanders grow rice, coconut, rubber, coffee, and tropical fruits and vegetables. Teak and structural timber are the islands' most important exports.

Jammu and Kashmir

Jammu and Kashmir is situated on the northernmost tip of India. Pakistan lies on the area's western border, and China lies to the east. Towering mountains surround the famous Vale of Kashmir, a fertile river valley watered by the Jhelum.

Jammu and Kashmir is a state in the far north of India. It is made up of three territories—Jammu, Kashmir, and Ladakh.

Jammu has a largely Hindu population. Kashmir is predominantly Muslim. Most Ladakhis are Buddhists.

Jammu and Kashmir is a land of towering mountain peaks, dense forests, deep blue lakes, and well-watered river valleys. Since ancient times, people have visited Kashmir to enjoy its natural charm and breathtaking scenery. Its summer capital, Srinagar, is known as the *Venice of the East* because it is crisscrossed by many canals. It was once the summer resort of India's Mogul rulers. The mountains of the Karakoram and Himalaya systems cut diagonally across the land. Two of the world's tallest mountain peaks are situated in Kashmir. K2 rises 28,250 feet (8,611 meters) in the Karakoram range. Nanga Parbat, whose name means *naked mountain,* rises 26,660 feet (8,126 meters) in the Pir Panjal range.

The Indus River separates the Himalaya and the Karakorum mountains as it flows northwestward through Kashmir. Its main tributary, the Jhelum River, flows through the famous Vale of Kashmir.

Jammu and Kashmir's heartland

The Vale of Kashmir extends about 85 miles (140 kilometers) from northwest to southeast. It is about 20 miles (32 kilometers) wide. The surrounding mountains protect the vale from the worst of the summer monsoons. As a result, the area enjoys moderate rainfall and long, warm summers. Winters are cold, with temperatures falling to 11° F. (−12° C).

The mild climate and well-watered soil of this river valley have encouraged the growth of agriculture. Rice is a major crop. It is grown on the plains where the paddies are watered by canals and on irrigated terraces on the lower mountain slopes. Corn and other grains are grown in the drier areas.

Saffron is another product of the vale. This food flavoring is obtained by drying the stigmas and part of the style of the purple

Aboard their *shikaras*, farmers and traders on Lake Dal bargain for produce. The flat-bottomed shikaras are used to transport passengers from shore to the houseboats that float on the lake.

autumn crocus. There are many orchards and vineyards as well. Roses and jasmine provide oil used in perfumes. A unique feature of the agriculture is the vegetable gardens floating on Kashmir's many lakes.

Large flocks of sheep and Cashmere goats graze on the high meadowlands of the vale. Underneath their top coat of hair, Cashmere goats have soft fleece that is used to produce cashmere, a fine silky wool.

A troubled paradise

Since 1947, Kashmir has been the center of a dispute between India and Pakistan. A truce line now divides Jammu and Kashmir into Indian and Pakistani sections. Neither country recognizes the jurisdiction of the other.

In the late 1980's, Muslims in the Indian section of Kashmir staged protests against Indian rule. Some demanded independence and some wanted Jammu and Kashmir to unite with Pakistan. In the early 1990's, Indian military forces clashed with protesters.

Shikaras, with canopied roofs and fluttering curtains, glide along one of Srinagar's many canals. The Jhelum River, spanned by seven wooden bridges, runs through this summer capital of Jammu and Kashmir.

The Vale of Kashmir, *above,* is a welcome and beautiful sight as drivers emerge from the Jawahar Tunnel in the Banihal Pass. Built in the 1970's at a height of about 9,000 feet (2,750 meters), the tunnel allows year-round travel through the mountain pass.

Farms are perched on the sides of hills in a remote river valley in Kashmir. Rice and corn are the main crops but only 6 per cent of the land is suitable for farming. Cedar and pine are important resources from Kashmir's dense forests.

Many protesters and some troops were killed. The tension and violence led to large-scale migration of Hindus from the region. In 1997 India and Pakistan pledged to reconcile their differences over the disputed territory. In October of that year India began to gradually withdraw its army from some towns in the Kashmir valley. However, nuclear tests conducted by India in 1998 intensified tensions over Kashmir.

643

Indonesia

The country of Indonesia consists of more than 13,600 islands that extend over 3,000 miles (4,800 kilometers) along the equator. Many of Indonesia's islands cover less than 1 square mile (2.6 square kilometers). However, Indonesia also includes about half of New Guinea and three-fourths of Borneo—the second and third largest islands in the world.

Tropical rain forests with many valuable hardwood trees cover much of Indonesia's land area. Crocodiles, elephants, pythons, rhinoceroses, and tigers live in some of these forests. Indonesia's mountains include about 60 active volcanoes. Although several violent volcanic eruptions have killed many people, Indonesians still live and farm near the volcanoes where volcanic ash makes the soil extremely fertile.

Today, most Indonesians live in small farm villages and still follow many of the ancient ways of life. For example, Javanese villagers celebrate important personal or family events with a ceremonial feast called a *selametan*. They dedicate the various foods to spirits and combine Muslim prayers with spirit worship.

The first settlers are thought to have arrived in Indonesia from the mainland of Southeast Asia between 2500 and 500 B.C. They established trade among the islands and with the Asian mainland. However, scientists have found bones in Java of one of the earliest types of prehistoric human beings. Now known as *Java man,* this prehistoric human being may have lived in the area as long as 1-1/2 million years ago.

In early times, the region from India to Japan, including Indonesia, was known to Europeans as the Indies. When Christopher Columbus arrived in America, he was really looking for a westward sea route from Europe to the Indies. He hoped to find riches in the Indies and to establish a great city for trading the products of the East and the West.

By the 1500's and 1600's, Portugal, England, and the Netherlands struggled to gain control of Indonesian trade. By the late 1700's, the Dutch controlled the trade of much of the islands.

Today, Indonesia stands at the crossroads of three cultures—the Chinese, the Indian, and the Southeast Asian. Influences of these cultures can be seen in the architecture, cookery, music, and dance of the country.

Indonesia Today

Indonesia is a mix of old and new. While the majority of Indonesians live in small farm villages, those who live in cities lead a modern lifestyle. The capital city of Jakarta on the island of Java is a bustling city with many high-rise buildings and wide boulevards. Surabaya, on the northeast coast of Java, is a major industrial center with a large university.

The modern era in Indonesia really began in 1908 when the first nationalist organization was founded. In 1927, as the people's desire for greater political power and independence grew, a civil engineer named Sukarno founded the Indonesian Nationalist Party. In 1945, Sukarno and other nationalists declared Indonesia's independence. Sukarno became president of the Republic of Indonesia, and a constitution was soon established.

The Dutch tried to regain control of the nation and, from 1945 to 1949, the Dutch and Indonesians fought many bitter battles. Although the Dutch recaptured much territory, they were unable to defeat the Indonesians. In November 1949, under pressure from the United States and the United Nations, the Dutch agreed to grant independence to Indonesia.

Guided democracy

Many political parties took part in the new government, but there was little agreement on how to solve Indonesia's many economic and social problems. Parliamentary elections, held in 1955, failed to produce a majority party. Revolts against the government broke out in 1958, but army units from Java defeated all the rebels by 1961.

In 1960, Sukarno dissolved the elected parliament and appointed a new one. He called his system of government *guided democracy*. In 1963, Sukarno was declared president for life.

FACT BOX

INDONESIA

COUNTRY

Official name: Republik Indonesia (Republic of Indonesia)
Capital: Jakarta
Terrain: Mostly coastal lowlands; larger islands have interior mountains
Area: 741,100 sq. mi. (1,919,440 km²)

Climate: Tropical; hot, humid; more moderate in highlands
Main rivers: Mahakam, Musi, Kayan
Highest elevation: Puncak Jaya, 16,503 ft. (5,030 m)
Lowest elevation: Indian Ocean, sea level

GOVERNMENT

Form of government: Republic
Head of state: President
Head of government: President
Administrative areas: 23 propinsi-propinsi (provinces), 2 daerah-daerah istimewa (special regions), 1 daerah khusus ibukota (special capital city district)

Legislature: Dewan Perwakilan Rakyat (House of Representatives) with 500 members serving five-year terms
Court system: Mahkamah Agung (Supreme Court)
Armed forces: 298,000 troops

PEOPLE

Estimated 2002 population: 217,314,000
Population growth: 1.63%
Population density: 296 persons per sq. mi. (114 per km²)
Population distribution: 61% rural, 39% urban
Life expectancy in years:
Male: 66
Female: 70
Doctors per 1,000 people: 0.2
Percentage of age-appropriate population enrolled in the following educational levels:
Primary: 113*
Secondary: 56
Further: 11
Languages spoken:
Bahasa Indonesia (official)

Indonesia, an island country in Asia, is about 25 per cent as large as the continental United States. It stretches for 3,000 miles (4,800 kilometers) between the Indian and Pacific oceans.

The Sukarno government badly mismanaged Indonesia's economy, and the country became almost bankrupt. Sukarno spent huge sums on monuments and sports stadiums but neglected the development of Indonesia's natural resources. The country's export business decreased while its debts rose rapidly. Prices, too, went up at an extremely high rate. In 1966, pressure from the army and student groups forced Sukarno to transfer much of his political power to General Suharto, the army leader. In 1968, Suharto became Indonesia's second president and established a government known as the New Order. Violent riots in 1998 and a deepening economic crisis led to the ouster of Suharto. He resigned in May of that year. In October 1999, the People's Consultative Assembly chose Abdurrahman Wahid as president and Megawati Sukarnoputri, one of Sukaro's daughters, as vice president. Wahid headed the National Awakening Party and an alliance of other Islamic parties. In July 2001, a special session of the People's Consultative Assembly voted to remove Wahid from office. Megawati became the country's new president.

Struggle for independence

In 2002, East Timor, which Indonesia annexed in 1976, became an independent country. Also in 2002, Indonesia gave Irian Jaya greater control over local affairs and renamed the province Papua. In December 2002, Indonesia's government reached a peace settlement with separatist rebels in Aceh.

English
Dutch
Local dialects

Religions:
Muslim 87%
Christian 10%

Enrollment ratios compare the number of students enrolled to the population which, by age, should be enrolled. A ratio higher than 100 indicates that students older or younger than the typical age range are also enrolled.

ECONOMY

Currency: Indonesian rupiah

Gross national income (GNI) in 2000: $119.9 billion U.S.

Real annual growth rate (1999–2000): 4.8%

GNI per capita (2000): $570 U.S.

Balance of payments (2000): $7,986 million U.S.

Goods exported: Oil and gas, plywood, textiles, rubber

Goods imported: Machinery and equipment, chemicals, fuels, foodstuffs

Trading partners: Japan, United States, Singapore, European Union

TECHNOLOGY

Radios per 1,000 people: 157

Televisions per 1,000 people: 149

Computers per 1,000 people: 9.9

Environment

Indonesia has a hot, humid climate. The lowlands have an average temperature of about 80° F. (27° C), but temperatures are lower in the highlands. Average temperatures vary little throughout the year. Java and the Lesser Sunda Islands are the only parts of the country with a dry season.

The islands of Indonesia can be divided into three groups: the Greater Sunda Islands, where most Indonesians live and where most economic activity is centered; the Lesser Sunda Islands, one of the few areas with a dry season; and the Moluccas, once known as the Spice Islands. Indonesia also includes Papua, which is part of New Guinea.

The Greater Sunda Islands

This island chain includes Borneo, Sulawesi, Java, and Sumatra. Of the four islands, Java has the most people and the largest city, Jakarta.

Borneo, the largest island in the Greater Sundas, has thick tropical rain forests and mountains in most of its interior. The southern part of Borneo—about three-fourths of the island—belongs to Indonesia. The area is thinly populated, and most people live along the coast. The Kapuas River, the longest river in Indonesia, flows about 700 miles (1,100 kilometers) from the mountains to the sea.

Forested Sulawesi (formerly Celebes), to the east of Borneo, is Indonesia's most mountainous island. Volcanoes, some of them active, rise on the northern peninsula. Some of Sulawesi's inland valleys and plateaus have fertile farmlands and rich grazing lands, while the coastal waters provide a bountiful catch.

Java, the most industrialized island of Indonesia, has thousands of small farm villages. An east-west chain of mountains extending across Java includes many volcanoes, some of them active. Wide, fertile plains lie north of the mountains, with limestone ridges to the south. A large highland plateau covers western Java.

Sumatra, to the west of Java, is the sixth largest island in the world. Along the southwestern coast of the island, a range of volcanic peaks rises about 12,000 feet (3,660 meters). The mountains slope eastward to a broad plain covered by dense rain forests and some farmland. Much of the eastern coast is swampy.

The Lesser Sunda Islands

This island chain extends eastward from Bali about 700 miles (1,100 kilometers) to Timor, the largest of the Lesser Sundas. Among these islands, Bali has the largest population and the largest city, Denpasar.

The Lesser Sundas have many mountains, including Mount Rinjani—the tallest peak—which rises 12,224 feet (3,726 meters) on the island of Lombok. Many small rivers flow from the mountains to the sea. The eastern islands have fewer rain forests and more dry grasslands than those in the west. Corn is the main crop in the east, while rice ranks first in the west. Most towns are coastal trading centers.

The Moluccas

The Moluccas lie between Sulawesi and New Guinea on both sides of the equator. Scattered among the large islands of the chain are hundreds of coral islands and *atolls,* groups of islands that enclose or partially enclose a lagoon. Most of these small

Lake Toba lies in the north central region of Sumatra. It is among the world's largest *calderas,* which are deep, caldronlike cavities on the summit of extinct volcanoes. Its elongated shape was produced by geological movement.

Mount Bromo, *above right,* in eastern Java, is one of a chain of volcanoes that crosses the major islands of Indonesia. The country has about 60 active volcanoes.

islands are unpopulated. Almost all of the large islands of the Moluccas are mountainous and thickly forested.

The people of the Moluccas live primarily in coastal trading settlements. Ambon, an important port, is the largest city. The Moluccas became famous hundreds of years ago when traders gathered spices there for sale in Europe. They were known as the Spice Islands.

Papua

Papua, the name given to western New Guinea, is a province of Indonesia formerly known as Irian Jaya. This area, the least developed and most thinly populated part of Indonesia, is a region of tropical rain forests and towering, snow-capped mountains. Puncak Jaya, the highest mountain in Indonesia, rises 16,503 feet (5,030 meters) over Papua. Cities, towns, and most farmland lie along the low, swampy coasts.

The volcano on the island of Krakatoa, which lies between Sumatra and Java, erupted in 1883 (A), causing one of the world's worst disasters. Much of Krakatoa Island disappeared. Volcanic ash rose 50 miles (80 kilometers), and a tidal wave killed about 36,000 people. Afterward, only the tiny peak of Anak Krakatoa ("child" of Krakatoa) remained (B).

Farmers cultivate rice, *left*, on small farms and on many kinds of land in Indonesia. In mountainous Bali, farmers build terraces and *dikes* (dirt walls) to catch rainfall for growing rice, the basic food of Indonesians.

People

ost Indonesians are Malays whose ancestors came from the mainland of Southeast Asia. Scholars believe that these people began coming to the islands about 4,500 years ago. Arabs, Chinese, Papuans, and other peoples also live in Indonesia.

More than 250 languages are spoken in Indonesia, including Bahasa Indonesia, the country's official language. Most children learn the language of their region at home before starting school. Later, when they enter school, they learn Bahasa Indonesia.

At the beginning of the 21st century, an estimated 225,000,000 people lived in Indonesia. About three-fifths of the people live in Java, which has 2,600 persons per square mile (1,000 persons per square kilometer). Because Java is so crowded, the government has encouraged people to move to less populated islands. However, Java's population continues to increase rapidly.

Many Indonesians, especially those born in Java, have only one name. For example, Sukarno, the country's first president, and Suharto, who became president in the 1960's, have only one name.

Traditional life

The great majority of Indonesians live in small villages and farm for a living. Village headmen and other traditional leaders, such as religious teachers, control most of village life. These traditional leaders govern by a system of local customs that stress cooperation. The villagers often settle disputes and solve problems by holding an open discussion that continues until an agreement is reached.

Most rural Indonesian families live in houses that consist of a sleeping room and a large living room. The living room may also serve as the kitchen. Most traditional Indonesian houses stand on stilts, and families use the space underneath the houses for cattle stalls, chicken coops, or storage for tools and firewood.

Some Indonesian groups build *long houses* that may house as many as 100 people. These groups include the Dyak peoples in Borneo and some Papuan groups in Papua. Many Indonesians decorate their walls with beautifully carved wood panels.

Almost 90 per cent of Indonesia's people are Muslims, and about 8 per cent are Christians. However, many Indonesians still hold

Muslims, followers of the Islamic religion, *above,* make up nearly 90 per cent of the population of Indonesia. Muslim traders from Arabia and India were among the first people to bring Islam to the region.

A Toradja village on the island of Sulawesi is home to one of Indonesia's many small ethnic groups. Many of the Toradja houses and granaries are decorated with beautifully carved wood panels.

other traditional religious beliefs and combine worship of ancient ancestors and nature with Islam or Christianity.

Traditional Indonesian clothing for both men and women is a *sarong*. A sarong is a colorful skirt that consists of a long strip of cloth wrapped around the body.

Modern life

Less than 10 per cent of Indonesia's people could read and write in 1945, when the government launched literacy programs conducted mainly in the villages. Today, about 84 per cent of the people 15 years old and older can read and write. Indonesian law requires children to attend elementary school for six years, beginning at least by age 8.

Many Indonesian dishes, *above,* are flavored with the spices that once made the islands famous. Rice, the country's chief food crop, is eaten with meat, fish, vegetables, or hot spices. Indonesian food is often cooked in coconut milk and oil and served on banana or coconut leaves.

Currently, nearly all of the country's children attend school.

Recreation in Indonesia has both traditional and modern influences. Indonesians in cattle-breeding areas hold ox races and bullfights. In some areas, people enjoy *Pencak silat,* an activity that combines dancing with self-defense.

Indonesians also enjoy several Western sports, including badminton, basketball, and soccer. The country's teams have won several world badminton championships. Many sports events are held in Jakarta's 200,000-seat stadium.

Jakarta, Indonesia's capital, is also the nation's most modern city. High-rise hotels and modern office and government buildings dominate the city's center, while major industries operate on the outskirts. Cars, buses, taxis, trucks, and motorcycles, as well as an electric railroad and two airports, serve the people of Jakarta. The capital also has several universities, including the University of Indonesia.

In Jakarta's fashionable residential section, many wealthy people live in large, expensive houses built by the Dutch when they ruled the country. Most of the people of Jakarta, however, live in small wood or bamboo structures without clean water, sewers, or electricity.

This Balinese boy's Western clothing contrasts sharply with the island's traditional way of life.

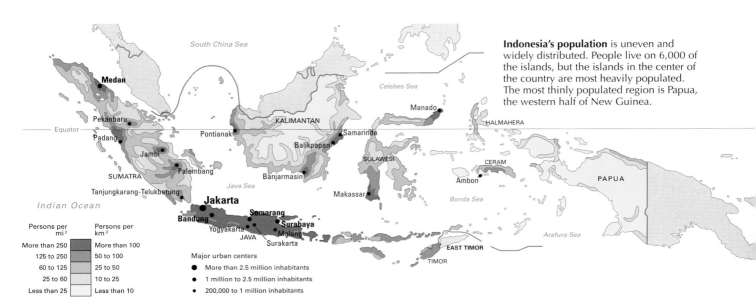

Indonesia's population is uneven and widely distributed. People live on 6,000 of the islands, but the islands in the center of the country are most heavily populated. The most thinly populated region is Papua, the western half of New Guinea.

South China Sea

Medan

Celebes Sea

Manado

HALMAHERA

KALIMANTAN

Equator — Pekanbaru

Padang

Pontianak

Samarinda

Balikpapan

SULAWESI

CERAM

Jambi

PAPUA

Palembang

SUMATRA

Banjarmasin

Ambon

Java Sea

Banda Sea

Tanjungkarang-Telukbetung

Makassar

Indian Ocean

Jakarta

Semarang

Bandung

Surabaya

Yogyakarta

Malang

Arafura Sea

JAVA

Surakarta

Persons per mi²	Persons per km²
More than 250	More than 100
125 to 250	50 to 100
60 to 125	25 to 50
25 to 60	10 to 25
Less than 25	Less than 10

EAST TIMOR

TIMOR

Major urban centers

● More than 2.5 million inhabitants

• 1 million to 2.5 million inhabitants

• 200,000 to 1 million inhabitants

Art and Culture

The Indonesian islands were a crossroads in the trade that extended from Arabia to China. Merchants of many lands, including Arabs, Chinese, Indians, and Persians, traded such goods as porcelain, textiles, and raw silk for Indonesian spices and sweet-smelling woods. Asian goods were also carried to Europe. Marco Polo, an Italian traveler, visited the islands in 1292.

The cultural contact that accompanied trade with the people of China, India, and the Arab world left an indelible mark on the Indonesian islands. The Buddhist, Hindu, and Islamic cultures in particular had a lasting effect on the religion, customs, and arts of the area.

Early influences

The ancestors of most Indonesians arrived on the islands between 2500 and 500 B.C. These ancient people established trade among the islands and with merchants on the Asian mainland.

Beginning in the A.D. 400's, Hinduism and Buddhism began to have a strong influence in Indonesia. Indian architecture influenced the design of Indonesian temples, and Indian legends became part of local puppet plays in the villages. Although villagers continued to worship according to their ancient beliefs, they also began to pray to Indian gods. Hindu and Buddhist kingdoms battled for power in Indonesia for hundreds of years. Many ancient temples from both religions remain standing, but few Indonesians today are Buddhists or Hindus.

The Muslim impact

Traders from Arabia and India were among the first people to bring the faith of Islam to Indonesia, but the greatest influence came from events in Melaka. A port kingdom on the Malay Peninsula, Melaka became a major trading power during the 1400's. When Melaka's ruler converted to Islam, the religion began to spread throughout the islands.

Today, the vast majority of the Indonesian people are Muslims, but many are less strict in their observance of Islam teachings than are Muslims in Arab countries. Indonesian Muslims have created their own version of the religion, combining ancestor and nature worship with Islam.

Literature, dance, drama, and crafts

Indonesia's arts include elements of the Hindu, Buddhist, and Islamic cultures. Ancient Hindu stories, for example, provide the basis for many of Bali's dramatic folk dances. The dancers wear elaborate costumes and headdresses and move to forceful rhythms.

Shadow-puppet dramas are also popular in Java and Bali. In these performances, which often last from 9 p.m. until 6 a.m., the shadows of the puppets are projected onto a white screen by a palm-oil lamp. Puppet plays based on Hindu myths are the most

Orchestras called *gamelans,* *below,* play an important part in Indonesian culture. The many gongs and drums of the gamelan produce a highly rhythmic sound. This music accompanies dance and shadow-puppet performances, sometimes lasting all night.

An Indonesian dancer is made up to look like a fierce animal. Many regional dances imitate the movements and habits of animals.

Traditional ceremonial marches, *right,* are popular in Indonesia. Every member of the family helps carry the elegant banners.

The legong dance, *above,* a Balinese folk dance, forms part of the anniversary rites of a village temple. The popular dance, usually performed by two or three girls, tells an ancient story of love and battle.

popular form of theater in Indonesia. An orchestra called a *gamelan* accompanies the dances and puppet plays. The gamelan has a distinctive sound created by metal gongs, flutes, xylophone-like instruments, double-ended drums, and a stringed instrument played like a cello.

In Indonesia's literature and crafts, as well, regional cultures mix with the Buddhist, Hindu, and Islamic. Early Indonesian literature, for example, consisted largely of local folk tales and traditional Hindu and Islamic stories. Indonesian sculptors worked in stone, creating beautiful decorations for their many ancient Hindu and Buddhist temples. The Balinese still carve Hindu figures and symbols for their homes and temples.

Economy

Indonesia produces only a small part of the manufactured goods the country needs. As a result, the nation relies heavily on foreign manufactured goods. On the other hand, Indonesia is a fertile land, rich in natural resources. These resources serve as the basis of the nation's economy. Farmers raise such crops as coffee, palm oil, rice, rubber, sugar cane, tea, and tobacco for export. Some farmers also raise water buffalo, cattle, goats, and poultry.

Rice and rubber

Rice, the chief food crop of Indonesia, is boiled or fried in a variety of ways and served with many other foods. It is grown on small farms, which also produce bananas, cassava, coconuts, corn, peanuts, spices, and sweet potatoes.

No one knows exactly when or where rice originated, but it probably grew wild and was gathered and eaten by people in Southeast Asia thousands of years ago. Today, Indonesia is one of the world's top rice-growing countries. Only China and India produce more rice.

Farmers in Java grow most of Indonesia's rice, producing at least two rice crops a year. Banks of earth are piled around the flat land to form fields called *paddies,* which are irrigated by water from mountain streams. In hilly areas, farmers build terraces and dirt walls to catch rainfall.

Most rice farming in Asia is done by hand. Farmers transplant seedlings from beds into the paddies. At harvest, farmers cut the rice stalks with knives or sickles and tie the stalks into bundles to dry. The dried stalks are then beaten against handmade screens to separate the grain.

Indonesia ranks second in the world in natural rubber production, with only Malaysia producing more. Almost all natural rubber comes from huge rubber tree plantations. Rubber trees grow best in hot, moist climates in acid, well-drained soils. The world's finest rubber-growing regions lie within a *rubber belt* that extends about 700 miles (1,100 kilometers) on each side of the equator.

A Balinese farmer takes his ducks to market. Raising ducks, which requires little space, is a good way to make a living on the crowded island of Bali.

Coconut palms are grown on small farms all over Indonesia, *below.* The people use coconut as a flavoring in cooking and often serve food on coconut palm leaves. They also use the leaves to weave baskets and to thatch roofs.

Rubber is one of the world's important raw materials. People depend so much on the special characteristics of rubber that it would be almost impossible to get along without it. Rubber is the only natural substance that is elastic, airtight, water resistant, shock absorbing, and long wearing.

Petroleum and timber

Petroleum and natural gas are Indonesia's chief exports. Indonesia is one of the chief producers of petroleum in the Far East and the only Southeast Asian member of the Organization of Petroleum Exporting Countries (OPEC), an association of nations that depend heavily on oil exports for their incomes. The members of OPEC work together to try to increase their revenue from the sale of oil on the world market. Two-thirds of the country's petroleum comes from the western island of Sumatra.

The country ranks seventh among the world's producers of forest products. Timber comes mainly from Borneo and Sumatra. Poor inland transportation on most of the other large islands of Indonesia interferes with the development of lumbering there. Indonesia's valuable hardwoods include ebony and teak, which are known for their beautiful grain patterns and are widely used in cabinets, flooring, furniture, and paneling.

Indonesia also has one of the world's largest fishing industries. The people take from the sea such fish as anchovies, mackerel, sardines, scad, and tuna. They catch milkfish and prawns in coastal ponds.

In 1997 and 1998, Indonesia suffered one of the worst financial slumps in its history. The value of its currency fell, and its stock market plunged. Banks and other businesses failed, and millions of people lost their jobs. Meanwhile, the price of food and other necessities soared. The crisis led to violence and to calls for Suharto's resignation. In May 1998, Suharto resigned and Vice President Bacharuddin Jusuf Habibie succeeded him.

Indonesia's paddies produce the population's main crop and staple food. Indonesian rice is grown mostly on small farms in irrigated fields, which are drained when the grain begins to ripen.

Woodcarvers rest after completing a year's work on a complicated panel. Some Indonesian peoples carve seated wooden figures to represent their ancestors and then pray to them. The Balinese carve Hindu figures and symbols for their homes and temples.

Rain Forest Under Threat

A tropical rain forest is a dense forest of tall trees in a region of year-round warmth and plentiful rainfall. Almost all such forests lie near the equator. Significant areas of both Malaysia and Indonesia are covered by these great forests, which take up large regions in Asia.

About half the world's species of plants and animals live in tropical rain forests. A rain forest has more kinds of trees than any other area in the world, with as many as 100 species in 1 square mile (2.6 square kilometers) of land. More species of amphibians, birds, insects, mammals, and reptiles live in these forests than anywhere else on earth.

The loss of the rain forest

Today, the rapid growth of the world's population and the increasing demands for natural resources threaten many tropical rain forests. People have destroyed large rain forests by clearing land for farms and cities. Mining, ranching, and timber projects have also caused great damage. This destruction of forests is called *deforestation*.

Until the late 1940's, tropical rain forests covered about 6 million square miles (16 million square kilometers) of the earth's land. In the late 1980's, they covered only about 4 million square miles (10 million square kilometers). About 50 million acres (20 million hectares) of tropical rain forests are destroyed each year, mainly in Latin America and Southeast Asia.

Deforestation has many far-reaching effects. The trees and other plants in forests help preserve the balance of gases in the atmosphere through a process called *photosynthesis*. In photosynthesis, the process plants use to make food, plants take in sunlight, carbon dioxide, and water, and give off oxygen. Renewal of the oxygen supply is vital to the continuing survival of oxygen-breathing organisms, such as animals and people. The absorption of carbon dioxide by plants is also important. An increase in carbon dioxide in the atmosphere could severely alter the earth's climate.

Forests also serve the earth by soaking up large amounts of rainfall. They prevent the rapid runoff of water, which may cause erosion and flooding. The trees and other plants

Lush rain forests, *above,* contain an incredible variety of plant and animal life. The tallest trees of a rain forest may reach a height of 200 feet (61 meters). A tropical rain forest is always green.

Deforestation, the destruction of forests, endangers the hundreds of species of plants and animals that live in Southeast Asian rain forests. Still, many of Southeast Asia's people clear forests to earn a living as farmers or timber workers.

Indonesia's tropical rain forests once covered most of the islands. However, the agricultural and industrial development of the 1900's has threatened to exhaust this vast resource. Rain forests around the world face a similar threat.

Planks of valuable hardwood, *below,* await shipment at a dockside in Jakarta. The export and sale of timber is vital to Indonesia's economy. The country ranks seventh among the world's producers of forest products.

The siamang, the largest of the gibbons, lives in the forests of Indonesia and Malaysia. The number of apes is decreasing because human activity has destroyed much of their habitat.

in forests also provide habitats for many living creatures. Scientists fear that further deforestation will lead to the extinction of hundreds of thousands of species of plants and animals.

The conservation predicament

Efforts at conservation, which involves the protection and wise use of the world's natural resources, face serious challenges today. One of the most difficult challenges is the need to find a compromise between protecting the environment and increasing—or even maintaining—agricultural and industrial production. Many Asian countries, for example, are hard-pressed to conserve their natural resources because the land must support so many people.

Southeast Asia has an average of 266 persons per square mile (103 per square kilometer)—about three times the world average population. Many of the region's forests have been cut down to produce timber and to clear land for farms and industries. In order to provide living and working space for people, it has been necessary to reduce the living space of wildlife.

Because every species plays a role in maintaining the balanced living systems of the earth, the loss of any species can threaten the survival of all life—including human beings. The goals of conservation can only be achieved through the combined efforts of the world's people.

Iran

Iran is located in Southwest Asia, north of the Persian Gulf. It lies south of the Caspian Sea, east of Iraq and Turkey, and west of Afghanistan and Pakistan.

Iran is one of the oldest countries on earth. Its history dates back almost 5,000 years to the first settlements of the Elamites in the southwestern region of present-day Iran. Later, the great Persian Empire included what is now Iran, as well as most of Southwest Asia and parts of Europe and Africa.

Throughout its history, Iran has been the site of many invasions and conquests. One of the most important was the Arab invasion of the 600's A.D. During this period, the Arabs converted the Persians to the religion of Islam. Today, most Iranians belong to the Shiah branch of Islam.

The discovery of oil in southwestern Iran during the early 1900's was another important development in Iran's history. Profits from oil exports promised to bring great wealth to the country. Reza Shah Pahlavi, who ruled Iran from 1925 to 1941, used these profits to modernize the nation and promote economic and social development.

His son, Mohammad Reza Pahlavi, became *shah* (king) in 1941 and introduced more economic and social reforms. The shah also used Iran's increasing oil revenues to develop many new industrial projects and provide a base for economic growth.

Many traditional Muslims disagreed with the shah's reforms. They believed that these reforms violated traditional Islamic teachings. In the late 1970's, opponents of the shah, led by the religious leader Ayatollah Ruhollah Khomeini, took control of the government and declared Iran an Islamic republic.

Since the revolution of 1979, Islamic tradition has replaced the Western ways introduced by the shah. Friday prayers at Teheran University are an example of the nation's return to Islamic customs.

Iran Today

The Islamic revolution of 1979 brought many changes to Iran. The government eliminated many Western influences and stopped almost all modernization. These political changes, along with the war with Iraq during the 1980's, weakened the country's economy.

Today, Iran faces many political and economic problems. The government is still troubled by fighting among its leaders and by acts of violence carried out by groups who oppose the government. Oil exports dropped sharply during the war with Iraq, and Iran was unable to pay for imports of food and other basic goods. Rising prices and high unemployment have added to Iran's problems.

Government

Since Iran's new Constitution was adopted in 1979, the country's supreme leader is the *faqih,* a scholar in Islamic law and the recognized religious leader of most Iranians. Ayatollah Khomeini was Iran's first faqih.

Iran's lawmaking body, the Majlis, is made up of 290 members elected to four-year terms. A Council of Guardians, consisting of 6 lawyers and 6 scholars, examines all proposed laws to make sure they do not violate the Constitution or Islamic principles.

Way of life

Islamic tradition is the basis of life in Iran. The Iranian government requires schools to teach Islamic principles. In the cities, women must wear a *chador,* a black, full-length body veil. The wearing of this veil is based on Islamic moral teachings.

All entertainment thought to be against the teachings of Islam is banned by the government. Freedom of speech and other civil rights have also been severely restricted.

After the revolution

The new government was bitterly anti-American because the United States had supported the shah. In October 1979, U.S. President Jimmy Carter allowed the shah to enter the United States for medical treatment. On November 4, Iranian revolutionaries seized the U.S. Embassy in Tehran. They held hostages for nearly two years while demanding the return of the shah for trial.

In 1980, Iran began fighting a war with Iraq. Hundreds of thousands of Iranians were killed or injured, and over a million people were left homeless. Since the cease-fire in 1988, Iran has taken part in negotiations for a peace treaty and has been trying to rebuild itself.

Ayatollah Khomeini died in 1989. Iran's top religious leaders chose President Ali Khamenei to succeed Khomeini as faqih.

Iran has been facing serious economic problems. The nation's oil exports have been

FACT BOX

COUNTRY

Official name: Jomhuri-ye Eslami-ye Iran (Islamic Republic of Iran)
Capital: Tehran
Terrain: Rugged, mountainous rim; high, central basin with deserts, mountains; small, discontinuous plains along both coasts
Area: 636,296 sq. mi. (1,648,000 million km²)

Climate: Mostly arid or semiarid, subtropical along Caspian coast
Main rivers: Karkeh, Zahreh, Mand, Atrek
Highest elevation: Mount Damavand, 18,386 ft. (5,604 m)
Lowest elevation: Caspian Sea, 92 ft. (28 m) below sea level

GOVERNMENT

Form of government: Theocratic republic
Head of state: Leader of the Islamic Revolution
Head of government: President
Administrative areas: 28 ostanha (provinces)

Legislature: Majles-e-Shura-ye-Eslami (Islamic Consultative Assembly) with 290 members serving four-year terms
Court system: Supreme Court
Armed forces: 545,600 troops

PEOPLE

Estimated 2002 population: 69,049,000
Population growth: 0.83%
Population density: 110 persons per sq. mi. (42 per km²)
Population distribution: 63% urban, 37% rural
Life expectancy in years: Male: 68 Female: 71
Doctors per 1,000 people: 0.9
Percentage of age-appropriate population enrolled in the following educational levels: Primary: 98 Secondary: 77 Further: 18
Languages spoken: Persian and Persian dialects 58%

IRAN

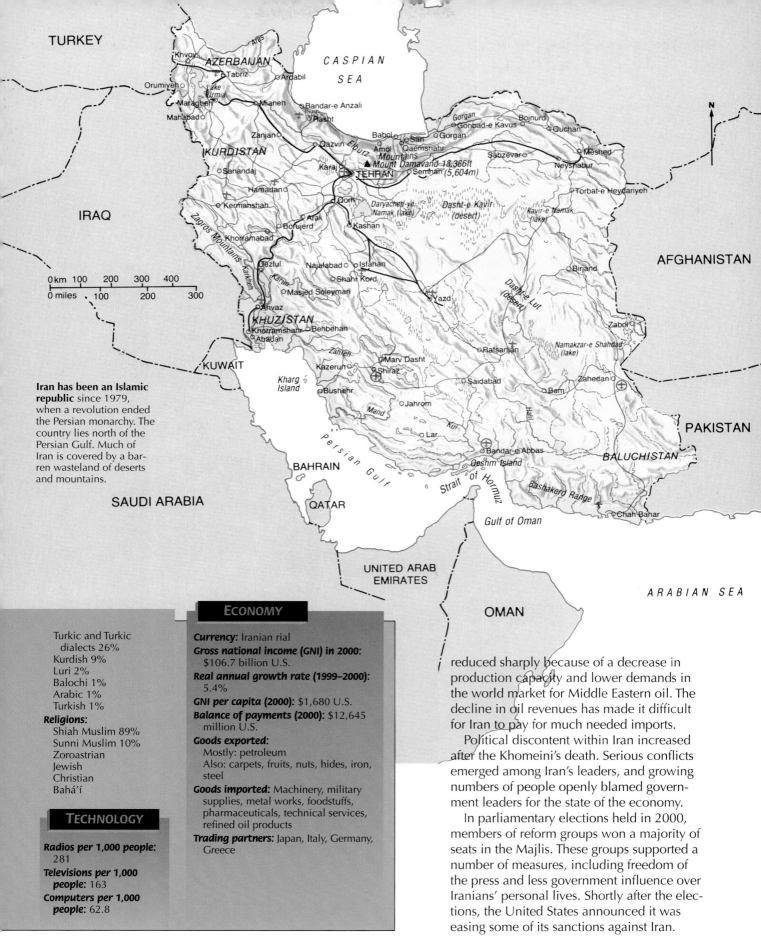

TURKEY

AZERBAIJAN

CASPIAN SEA

Khvoy
Orumiyeh
Tabriz
Ardabil
Maragheh
Lake Urmia
Mahabad
Mianeh
Bandar-e Anzali
Rasht
Gorgan
Gonbad-e Kavus
Bojnurd
Quchan

KURDISTAN

Zanjan
Qazvin
Babol
Sari
Gorgan
Sabzevar
Meshed
Neyshabur

IRAQ

Sanandaj
Karaj
TEHRAN
Elburz Mountains
Mount Damavand 18,386ft
Semnan (5,604m)
Torbat-e Heydariyeh

Hamadan
Kermanshah
Qom
Daryacheh-ye Namak (lake)
Dasht-e Kavir (desert)
Kavir-e Namak (lake)

AFGHANISTAN

Arak
Borujerd
Kashan
Khorramabad

Zagros Mountains
Karkheh
Dezful
Karun
Najafabad
Isfahan
Shahr Kord
Masjed Soleyman
Yazd
Birjand

Dasht-e Lut (desert)

KHUZISTAN
Ahvaz
Khorramshahr
Abadan
Behbehan
Zahreh
Marv Dasht
Kazerun
Shiraz
Rafsanjan
Saidabad
Bam
Zahedan
Namakzar-e Shahdad (lake)
Zabol

KUWAIT

Kharg Island
Bushehr
Jahrom
Mand
Lar
Kul
Saidabad

BALUCHISTAN
Bashakerd Range

Bandar-e Abbas
Qeshm Island
Strait of Hormuz
Hali

PAKISTAN

Iran has been an Islamic republic since 1979, when a revolution ended the Persian monarchy. The country lies north of the Persian Gulf. Much of Iran is covered by a barren wasteland of deserts and mountains.

Persian Gulf

BAHRAIN

SAUDI ARABIA

QATAR

UNITED ARAB EMIRATES

OMAN

Gulf of Oman

Chah Bahar

ARABIAN SEA

0 km 100 200 300 400
0 miles 100 200 300

N

Turkic and Turkic
dialects 26%
Kurdish 9%
Luri 2%
Balochi 1%
Arabic 1%
Turkish 1%
Religions:
Shiah Muslim 89%
Sunni Muslim 10%
Zoroastrian
Jewish
Christian
Bahá'í

TECHNOLOGY

Radios per 1,000 people:
281
Televisions per 1,000 people: 163
Computers per 1,000 people: 62.8

ECONOMY

Currency: Iranian rial
Gross national income (GNI) in 2000: $106.7 billion U.S.
Real annual growth rate (1999–2000): 5.4%
GNI per capita (2000): $1,680 U.S.
Balance of payments (2000): $12,645 million U.S.
Goods exported:
Mostly: petroleum
Also: carpets, fruits, nuts, hides, iron, steel
Goods imported: Machinery, military supplies, metal works, foodstuffs, pharmaceuticals, technical services, refined oil products
Trading partners: Japan, Italy, Germany, Greece

reduced sharply because of a decrease in production capacity and lower demands in the world market for Middle Eastern oil. The decline in oil revenues has made it difficult for Iran to pay for much needed imports.

Political discontent within Iran increased after the Khomeini's death. Serious conflicts emerged among Iran's leaders, and growing numbers of people openly blamed government leaders for the state of the economy.

In parliamentary elections held in 2000, members of reform groups won a majority of seats in the Majlis. These groups supported a number of measures, including freedom of the press and less government influence over Iranians' personal lives. Shortly after the elections, the United States announced it was easing some of its sanctions against Iran.

People

Iran stands at one of the world's major cross-roads, between the Near East and central Asia. Because of its location, people have migrated across the region for centuries. It has been the site of many civilizations and empires.

Ancient settlers

The first people to arrive in what is now Iran were the Elamites. They may have settled in southwestern Iran as early as 3000 B.C.

During the 1500's B.C., tribes of Aryans invaded Iran from central Asia. Today, about two-thirds of the Iranian population are descended from the Aryans. The name *Iran* itself comes from a Persian word meaning *Land of the Aryans.*

The modern Persian language, called Farsi, developed from the language of the Aryans. When the Arabs conquered Iran in the A.D. 600's, Farsi absorbed many Arabic words and Arabic script.

Ethnic groups

The Persians, Iran's largest ethnic group of Aryan origin, form about 60 per cent of the country's total population. The majority of Persians live in central Iran and on the slopes of the Elburz and Zagros mountains.

Other groups believed to be descended from the Aryans include the Gilanis and Mazandaranis of the north, the Kurds of the northwest, the Lurs and Bakhtiaris of the west, and the Baluchis of the southeast.

The Baluchis of southeast Iran belong to the Sunni, or orthodox, branch of Islam. Religious differences between the two branches of Islam have been a source of tension in the Baluchis's relations with the Iranian government. About half of the Baluchis are nomadic or partly nomadic. The other half are farmers and villagers. Because communication between the government and the Baluchi settlements is poor, these people are among the poorest and least educated in Iran.

Non-Aryan ethnic groups include the Azerbaijanis, the Khamseh, the Qashqais, the Turkomans, and Arabs. The Azerbaijanis are Shiah Muslims and make up about one-third of the population of Tehran. The Turkomans, a

Iranian women have less freedom than women in other Islamic countries. Many women in the cities wear traditional long body veils called *chadors.*

A fisherman displays his catch, *above,* along the Caspian coast, where most of Iran's people live.

A group of Bakhtiaris enjoys a typical meal. About 250,000 Bakhtiaris live in the Zagros Mountains in western Iran and still lead a nomadic life. Every year these herders lead their flocks up to the high summer pastures of Khuzistan. The Bakhtiaris's artistic skills can be seen in their colorful, boldly patterned carpets.

tribal people who live on the southeastern edge of the Caspian Sea, are also mainly Shiah Muslims.

Religion

About 99 per cent of the Iranian people are Muslims. About 89 per cent of them belong to the Shiah branch of Islam, which is Iran's state religion. Although the Islamic government has little tolerance for other religions, a few religious minorities do exist.

The largest religious minority are the Bahá'ís. The Bahá'í faith grew out of the Bábi faith, a religion founded in Persia in 1844. Bahá'ís, who number about 300,000 in Iran, are forbidden by the government to practice their faith. They have been severely persecuted for their beliefs.

Iran also has some Christians, Jews, and followers of Zoroastrianism, an ancient Persian religion.

Iranian refugees

More than one million Iranians have left their country since the Islamic revolution and the start of the war with Iraq. Many found shelter in the neighboring countries of Pakistan and Turkey. Some fled to Western Europe and the United States.

Most of these refugees were middle-class Iranians who lived in the cities. They included technicians and scientists, as well as artists and teachers. The departure of these educated people left Iran in great need of skilled professionals.

Land and Economy

Iran lies in an arid zone, dominated by deserts and mountains. The country is, in fact, one of the most mountainous in the world.

There are no major river systems in Iran. Therefore, throughout history, people have traveled across the country by camel caravan. These caravans followed routes through gaps and passes in the mountains.

Landscape and climate

Iran's landscape can be divided into four regions: (1) the Mountains, (2) the Interior Plateau, (3) the Caspian Sea Coast, and (4) the Khuzistan Plain.

Iran has two mountain ranges, the Elburz and the Zagros. The Elburz Mountains stand along the northern border between the Caspian Sea Coast and the Interior Plateau. The northern slopes of the Elburz receive a great deal of rainfall, so the farmers are able to grow many crops in the region's rich soil. The southern slopes are barren and dry.

The Zagros Mountains begin at the borders of Turkey and Azerbaijan and stretch south and east to the Persian Gulf. The valleys of the northern and central part of the range are well populated. The dry, rugged southern sections are far less populated.

Winters are long and bitterly cold in the mountain regions. Temperatures can drop as low as −20° F. (−29° C). The summers are mild.

The Elburz and Zagros mountains surround the Interior Plateau of central and eastern Iran. The Interior Plateau stands about 3,000 feet (900 meters) above sea level and covers about half of Iran's area.

Most of the Interior Plateau consists of two deserts, the Dasht-e Kavir and the Dasht-e Lut. These deserts are among the most arid and barren on earth. Together, they cover more than 38,000 square miles (98,000 square kilometers).

Summer temperatures in the desert soar as high as 130° F. (54° C). Rainfall is sparse—only about 2 inches (5 centimeters) per year. Temperatures in the cities of the plateau are less extreme.

The Caspian Sea Coast is a narrow strip of lowland between the Caspian Sea and the

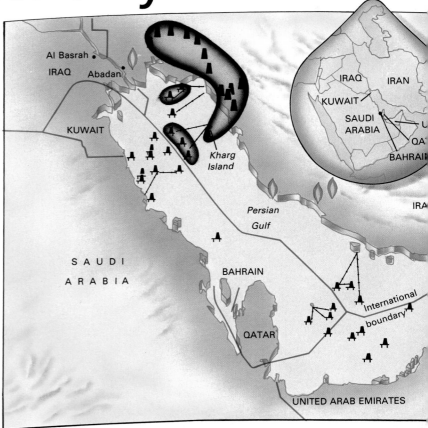

Most of Iran's oil fields lie in the southwest, near the city of Abadan. Some reserves also lie under the waters of the Persian Gulf. Iran is a leading oil producer.

Handwoven carpets, made by Iranian craftworkers, dry in the sun. Iranian rugs are prized all over the world for their graceful patterns and soft colors.

slopes of the Elburz Mountains. Because of its mild climate and plentiful rainfall, the Caspian Sea Coast is the most heavily populated region in Iran.

The Khuzistan Plain lies north of the Persian Gulf, between the border of Iraq and the Zagros Mountains. It is a very important region for Iran, because it contains the country's richest oil deposits. Many crops are also grown in the Khuzistan Plain's rich farmland. Summers on the plain are very hot and humid, but winters are mild and pleasant.

Bandar-e Abbas

Strait of Hormuz

Pipeline
Oilfield/
oil platform
Natural gas
field

0 km   100
0 miles   60

N

The economy

Due to the scarcity of water, only 12 per cent of Iran's land can be farmed. About 75 per cent of the cultivated land is used to grow wheat and barley. Other crops include corn, cotton, dates and other fruits, lentils, nuts, rice, sugar beets, tea, and tobacco.

Since it was first discovered 125 miles (200 kilometers) north of Abadan in 1908, oil has been the backbone of Iran's economy. With about 48 billion barrels of petroleum in the country's oil fields, Iran is one of the world's leading oil producers. Petroleum output has decreased sharply since the 1979 revolution, causing great damage to Iran's economy.

The oil refinery at Masjed Soleyman, *above left,* stands near Iran's first oil field. The country's oil production declined after the 1979 Islamic revolution.

The eggs of sturgeon caught in the Caspian Sea make a delicacy called *caviar.* Sturgeon eggs are the most important product of Iran's fishing industry.

Bakhtiari herders drive flocks of sheep and goats on their yearly migration over the Zagros Mountains. These nomads live in round tents made of black felt.

History

The history of Iran begins as far back as 3000 B.C., when an ancient people called the Elamites lived in what is now south-western Iran. About 1,500 years later, two groups of Aryans invaded the region from Central Asia. The first group set up the king-dom of Media in the northwest. The second group settled in the south of the region, later known as Persia.

In 550 B.C., the Persians, led by Cyrus the Great, overthrew the Medes and established the Achaemenid dynasty. The Achaemenids united Media and Persia, and by 539 B.C., they had added Babylonia, Syria, Palestine, and all of Asia Minor to their empire.

The golden age of Persia

In 522 B.C., Darius I became king of the Achaemenid Empire, which then included most of the known world. The empire prospered under his reign.

By the mid-400's B.C., the glory of the Achaemenid Empire was almost gone. In 331 B.C., Alexander the Great of Macedonia conquered the empire. After Alexander's death, one of his generals, Seleucus, founded a new dynasty, the Seleucid. The Seleucids governed Iran until about 250 B.C., when the Parthians defeated them. In A.D. 224, the Parthians were conquered by the Persians, who established the Sassanid dynasty.

The rise of Islam

The Arabs conquered the Sassanians in the mid-600's. The Arabs gradually converted most Iranians to Islam. Arabic replaced Persian as the official language of government in Iran, but most common people continued to speak Persian. Under the reign of the Arabs, Iran became a world center of art, literature, and science.

By the mid-1000's, Seljuk Turks had conquered most of Iran. In 1220, the Mongols, led by Genghis Khan, conquered Iran, destroying many cities and killing thousands of people.

In the late 1400's and early 1500's, the Sa-

Under the Sassanid dynasty, *above*, Iranian art flourished and Iran regained the glory of the ancient Achaemenid Empire.

c. 3000 B.C. The region that is now Iran is settled by Elamites.

c. 1500's B.C. Aryan settlers divide the region into Media and Persia.

550 B.C. Cyrus the Great overthrows Media and founds Achaemenid Empire.

539 B.C. Cyrus conquers Babylonia, Palestine, Syria, and Asia Minor.

331 B.C. Alexander the Great conquers Persia.
323–250 B.C. The Seleucids rule Iran.
250 B.C. Parthian armies conquer Iran.

A.D. 224 Persians overthrow Parthians and found Sassanid dynasty.

637 Muslim Arabs conquer Iran.

Mid 1000's–1220 Seljuk Turks rule Iran.
1220 Mongols led by Genghis Khan conquer Iran.
1501–1722 The Safavid dynasty rules Iran.

1736 Nadir Shah conquers Iran.
1794–1925 The Qajar dynasty rules Iran.

1826 Russia invades Iran.
1828 Treaty of Turkomanchai establishes boundaries.
1856–1857 Iran invades Afghanistan. Iran signs peace treaty with Great Britain.

Early 1900's Anglo-Persian Oil Company develops oil fields in Iran.
1906 Shah Muzaffar al-Din signs Iran's first Constitution.
1925 Reza Khan (Reza Shah Pahlavi) becomes shah.
1941 Mohammad Reza Pahlavi succeeds his father to the throne.
1951 Iran's oil industry is nationalized.
1979 Revolutionaries force shah out of Iran and establish Islamic republic.
1980 Outbreak of Gulf War with Iraq.
1987 Dispute with United States in Persian Gulf over safety of Kuwaiti oil tankers.
1988 Gulf War ends in cease-fire.
1989 Death of Ayatollah Khomeini, leader of the 1979 revolution and faqih of Iran.

Khosrau II, *left*, reigned from A.D. 590 to 628.

Mohammad Reza Pahlavi, *far left*, was shah from 1941 to 1979.

Ayatollah Ruhollah Khomeini, *left*, was faqih from 1979 to 1989.

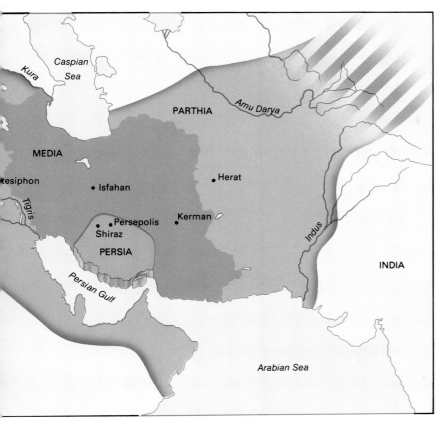

favids, a Turkish tribe, gained control over several regions of Iran. The Safavid kings ruled Iran until 1722, when armies from Afghanistan invaded the country.

Nadir Shah, a Turkish tribesman, drove the Afghans out of Iran in the 1730's and became king. After Nadir Shah was assassinated in 1747, civil war broke out in Iran, as many leaders fought for control of the country.

In the late 1700's the Qajar dynasty came to power and ruled Iran until 1925. Reza Khan became shah in 1925 after overthrowing the Qajar government. He then changed his family name to Pahlavi. Reza Shah tried to modernize Iran and eliminate foreign interference.

Soon after World War II began in 1939, Iran declared itself neutral. However, the Allies wanted to use the Trans-Iranian Railway to ship war supplies to the Soviet Union. Reza Shah objected, and in 1941, Soviet and British troops invaded Iran. They forced Reza Shah to give up his throne. His son, Mohammad Reza Pahlavi, became shah. The new shah signed a treaty with the Allies, allowing them to ship supplies through Iran.

During the early 1960's, the shah began a series of economic and social reforms. Like his father, Mohammad Reza Pahlavi wanted to modernize Iran.

The shah ruled with absolute control over the government. Many Iranians opposed his use of the Savak—a secret police force. They also believed that his modernization programs violated traditional Islamic teachings.

In January 1979, the shah was forced to leave the country. The following month, revolutionaries led by Ayatollah Ruhollah Khomeini took over the government and established Iran as an Islamic republic.

The royal mosque at Isfahan, built in 1628–1629, shows the influence of many different cultures on Iranian architecture.

Art and Architecture

Iran's artistic heritage begins with its earliest civilizations. Painted pottery dating back to about 5000–4000 B.C. has been found in western Iran.

From these beginnings, Iranian artists and architects have given the world some of its most splendid paintings, architecture, metalwork, ceramics, and textiles.

The Achaemenid Empire

When Cyrus the Great united Persia and Media under one empire in 550 B.C., he established the great Achaemenid Empire. These early invaders of ancient Persia expressed the glories of their rule through art and architecture.

The Achaemenid palaces were huge, impressive structures. The palace at Persepolis, the capital of the Achaemenid Empire, features a large number of rooms, halls, and courts set on a raised platform. The structure blends the influence of many different cultures to create a uniquely Persian style. Its 40-foot (12-meter) columns show the influence of Egyptian architecture and the ornamental detail at its bases came from the Ionian Greeks in Asia Minor.

Sassanid art and architecture

The Sassanid Empire was known for its metalwork. Goblets, plates, and other objects were made of gold and featured fine ornamental detail. Sassanian artists also carved enormous sculptures in rock.

The Sassanid Empire's chief glory, however, was its woven silk. Large amounts of Sassanian woven silk were exported to Constantinople and the Christian West. Its many colors and patterns influenced the art of the Middle Ages as well as Islamic art.

Islamic art

When the Arabs invaded Persia in the A.D. 600's, they brought with them the religion of Islam. The people who accepted Islam blended the influence of the highly developed art of Persia with the religious customs of their new-found faith. They developed a uniform style of art known as *Islamic art.*

Islamic art is noted for its mastery of technique, design, and color. Islamic artists sought to make all aspects of life beautiful. Examples of Islamic art can be seen in architecture, textiles, metalware, pottery, carved and molded plaster, glassware, wood and ivory carvings, and book *illuminations* (decorations). Islamic art reached its peak between 800 and 1700.

Islamic teachings forbid Muslims to create images of living things. Muslims believe that *Allah* (God) is the one and only creator of life. So Islamic artists developed a special type of decoration called *arabesque.* It consists of winding stems with abstract leaves. Islamic artists stylized living things in other ways to make them look like they weren't ac-

A Persian rug from the 1500's shows the typical dark colors and floral patterns. Islamic craft workers developed carpet weaving into a fine art.

tually alive. For example, they would often present figures incompletely, or make them appear faceless, or make them look like flattened shapes.

Another important development in Islamic art was the use of Arabic script, which lends itself to beautiful writing called *calligraphy*. Islamic artists often used calligraphy to decorate art objects and walls of religious buildings with writings from the *Koran*, the holy book of Islam.

Islamic artists found their greatest expression in architecture, especially *mosques*. Mosques are Islamic houses of worship. The city of Isfahan features one of Iran's most magnificent mosques—*Masjid-e-Shah* (Shah Mosque). The Mosque of Shaikh Lutfullah, with glazed tilework adorning its elaborate entrance and dome, is also in Isfahan.

A painting dating from the 1300's portrays a scene from Persia's most famous epic poem, the *Shahnama* (*Book of Kings*), written by Firdausi.

Columns that once supported a wooden ceiling over the Audience Hall of Darius, in about 500 B.C., still stand in Shiraz, the site of ancient Persepolis.

Iraq

Iraq is an Arab country situated at the head of the Persian Gulf. It is almost entirely landlocked. Iraq's only outlet to the Persian Gulf is through the Shatt al Arab, a waterway connecting the Tigris and Euphrates rivers. The country's most important port, Al Basrah, stands on the banks of the Shatt al Arab about 55 miles (90 kilometers) from the Persian Gulf.

Iraq's history began more than 5,000 years ago in the Tigris-Euphrates Valley. It was here that the Sumerians developed the world's first civilization. Sumer was an ancient region in southern Mesopotamia, which is now southeastern Iraq.

The Sumerian civilization began about 3500 B.C. Archaeologists have found that the Sumerians built magnificent palaces and temples. They also had knowledge of mathematics, astronomy, and medicine. The Sumerians invented the world's first writing system. The system began chiefly as a set of word pictures and developed into a script called *cuneiform*.

Built on the ruins of the world's first civilization, Iraq has developed into one of the top five oil-producing nations in the world. Its history has been long and, in recent years, quite eventful.

During the late 1900's, Iraq became involved in two bitter disputes. Between 1980 and 1988, war broke out between Iraq and Iran over territorial disputes and other disagreements. The war left Iraq with a damaged economy and many foreign debts.

In August 1990, Iraq invaded and occupied Kuwait. Many other countries condemned the invasion, and in January 1991, an international coalition led by the United States used military force to drive Iraq from Kuwait. This conflict became known as the *Persian Gulf War*.

Ancient empires

The Sumerians were the first civilization to live in Iraq. Their advanced society thrived there until about 2000 B.C. The Sumerian region centered around the great city of the ancient world, Babylon.

Archaeological records show the first mention of Babylon about 2200 B.C. Babylon was the capital of the kingdom of Babylonia and of two Babylonian empires. The present-

day Iraqi city of Al Hillah stands on the site of Babylon. It is about 60 miles (97 kilometers) south of present-day Baghdad, Iraq.

Babylon was famed for its developments in learning, architecture, sculpture, mathematics, and geometry. Its culture strongly influenced ancient Greek philosophers. The Hanging Gardens of Babylon are one of the Seven Wonders of the Ancient World. These gardens, laid out on brick terraces about 75 feet (23 meters) above the ground, were probably built by King Nebuchadnezzar II for his wife.

The Babylonian Empire lasted until about 1800 B.C., when the Assyrian Empire ruled the area. Assyria declined after the mid-600's B.C.

In 539 B.C., the Persians conquered the region and added it to their empire. When Alexander the Great defeated the Persians in 331 B.C., he began a period of Greek rule. The region became part of the Roman Empire in A.D. 115.

The Arabs spread through Mesopotamia in A.D. 637, bringing the Arabic language and the Islamic religion to the region. Baghdad was the capital of the Arab Empire, and it became a great center of learning. The town of Samarra, with its ruined Great Mosque, also served briefly as the Arab capital in the 800's.

During the 1200's, Baghdad was destroyed in a Mongol invasion of the region. This area never regained its power or importance after the Mongol conquest. It became part of the Ottoman Empire when the Ottoman Turks seized Mesopotamia in 1534.

Modern Iraq

The Turks ruled the Mesopotamian region for almost 400 years. During World War I, troops from Great Britain took Mesopotamia from the Ottoman Turks. In 1920, the League of Nations gave Great Britain administrative rule in Mesopotamia. The British helped the region's leaders set up a government in 1921. These leaders named the country Iraq and elected King Faisal I the first monarch.

Iraq gained its independence as a nation in 1932. During the following years, a series of military revolts ended the monarchy, and the revolutionaries declared Iraq a republic in 1958.

Iraq Today

The history of Iraq since the 1960's has been dominated by political turmoil and bloody conflicts with its neighbors in the Arab world. In 1963, the rebels who had established Iraq as a republic were overthrown by Abdul Salam Arif and Ahmed Hasan al-Bakr, who were members of the Baath Party.

The party's philosophy blends socialism and Arab nationalism. It forms the basis of education in Iraq from the primary grades through the university levels.

In 1968, al-Bakr set up a Baathist-controlled military government in Iraq. In 1979, al-Bakr resigned from office, and Saddam Hussein succeeded him.

War against Iran

In 1980, Iraq entered into a war against Iran, which proved to be very harmful to Iraq. War damage was heavy, even though most of the battles were fought on Iranian soil. Many Iraqi soldiers and civilians were killed. Missiles struck Baghdad and other cities. Iraq's normal trade routes were disrupted, and its ports were closed. In August 1988, both sides agreed to a cease-fire.

The Persian Gulf War of 1991

In August 1990, Iraqi forces under Hussein's leadership invaded and occupied Kuwait. Hussein claimed that Kuwait was violating oil production limits set by OPEC, thus lowering the price of oil. He also hoped that, by annexing Kuwait as an Iraqi province, Iraq could absorb the country's wealth. This would boost Iraq's economy, which had been badly damaged in the war with Iran.

However, the United Nations Security Council declared the annexation null and void. In November 1990, it approved the use of force to remove Iraqi troops from Kuwait if they did not leave Kuwait by Jan. 15, 1991. Iraq still refused to withdraw, and a coalition of international forces led by the United States attacked military and industrial targets in Iraq and Kuwait.

Later, allied forces launched a ground attack into Kuwait, quickly defeating Iraqi troops. Two days later, Iraqi troops began withdrawing from Kuwait.

An estimated 1,500 to 100,000 Iraqi soldiers died as a result of the war, and bombing severely damaged Iraq's transportation systems, communication systems, and many industries. The attacks also destroyed Iraq's ability to provide electric power and clean water to its cities and towns.

Recent developments

Iraq's economy suffered greatly after the Persian Gulf War. The United Nations (UN) imposed sanctions that banned Iraq from selling oil, its main revenue source. The sanctions caused serious economic problems, and Iraq

FACT BOX

COUNTRY

Official name: Al Jumhuriyah al Iraqiyah (Republic of Iraq)
Capital: Baghdad
Terrain: Mostly broad plains; reedy marshes along Iranian border in south with large flooded areas; mountains along borders with Iran and Turkey
Area: 168,754 sq. mi. (437,072 km²)

Climate: Mostly desert; mild to cool winters with dry, hot, cloudless summers; northern mountainous regions experience cold winters with occasional heavy snows that melt in early spring, sometimes causing extensive flooding in central and southern Iraq
Main river(s): Euphrates, Tigris, Little and Great Zab
Highest elevation: Haji Ibrahim, 11,811 ft. (3,600 m)
Lowest elevation: Persian Gulf, sea level

GOVERNMENT

Form of government: Republic
Head of state: President
Head of government: Prime minister
Administrative areas: 18 muhafazat (provinces)

Legislature: Majlis al-Watani (National Assembly) with 250 members serving four-year terms
Court system: Court of Cassation
Armed forces: 429,000 troops

PEOPLE

Estimated 2002 population: 24,451,000
Population growth: 2.86%
Population density: 144 persons per sq. mi. (56 per km²)
Population distribution: 68% urban, 32% rural
Life expectancy in years:
Male: 66
Female: 68
Doctors per 1,000 people: 0.5
Percentage of age-appropriate population enrolled in the following educational levels:
Primary: 88
Secondary: 20
Further: 13

Iraq, *left,* is an Arab republic lying at the head of the Persian Gulf. The country features a wide variety of landscapes. Stony deserts make up most of the southern regions, and mountains rise up north of the Tigris and Euphrates rivers.

In Baghdad, craftsmen work on brass pots. Iraqi metalworkers produce beautiful trays, pitchers, and other objects.

ECONOMY

Languages spoken:
Arabic
Kurdish (official in Kurdish regions)
Assyrian
Armenian

Religions:
Muslim 97%
(Shiah 60%-65%,
Sunni 32%-37%)
Christian

Currency: Iraqi dinar
Gross domestic product (GDP) in 2001: $59 billion U.S.
Real annual growth rate (2001): -5.7%
GDP per capita (2001): $2,500 U.S.
Balance of payments (2000): N/A
Goods exported: Crude oil
Goods imported: Food, medicine, manufactures
Trading partners: Russia, France, China, Egypt

TECHNOLOGY

Radios per 1,000 people: 222
Televisions per 1,000 people: 83
Computers per 1,000 people: N/A

said that hundreds of thousands of people had died of disease and malnutrition by the mid-1990's.

In November 1994, Iraq formally recognized the independence of Kuwait. Also in 1994, fighting broke out between rival groups of Kurds in northern Iraq. In 1996, the Iraqi government sent troops to support one of the Kurdish groups. The United States opposed this action and launched missiles against southern Iraq.

In November 1996, the UN partially lifted the embargo on Iraq by allowing it to export oil on a limited basis. In 1997, Hussein ordered all United States personnel affiliated with UN inspections of Iraqi weapons programs out of the country. In December 1998, U.S. and British forces launched four days of air and missile strikes against Iraq.

In November 2002, the UN Security Council passed a resolution demanding the resumption of weapons inspections in Iraq. In March 2003, U. S. forces launched an air attack against Baghdad. The military campaign intended to overthrow Hussein and eliminate Iraq's ability to produce weapons of mass destruction.

Environment

Iraq, which covers an area slightly larger than the state of California, can be divided into four land regions. They are (1) the upper plain, (2) the lower plain, (3) the mountains, and (4) the desert.

The upper plain

The upper plain is made up of the region that lies between the Tigris and Euphrates rivers north of the city of Samarra. The upper plain is mostly dry, rolling grassland. The highest hills in the region rise about 1,000 feet (300 meters) above sea level. The Tigris and Euphrates flow swiftly through the upper plain, carrying large amounts of sediment.

Mosul is the largest city in the upper plain, with about 1 million inhabitants. It is located in the northern sector of the region, about 60 miles (100 kilometers) from the Syrian and Turkish borders. Since 1952, Mosul has been the center of an important oil field.

The lower plain

The lower plain extends from Samarra to the Persian Gulf. It includes the delta between the Tigris and the Euphrates. Although the lower plain receives little rainfall, the farmlands are green with vegetation. An elaborate system of irrigation canals, first dug centuries ago, brings water to the crops.

Farmers also rely on regular flooding of the region to water their crops. Floods occur in April and May each year, when the Tigris and Euphrates overflow their banks.

Old-fashioned water wheels still stand on the river banks. Here, palm, orange, and lemon groves have provided abundant crops of fruit for centuries. In recent years, modern concrete canals have replaced some of the old systems.

More advanced methods of irrigation allow Iraqi farmers to extend their fields into the drier areas. As a result, they now have enough water to grow wheat, cotton, rice, tomatoes, oranges, melons, and grapes.

South of Baghdad lie huge groves of date palms. This region is renowned for its tall, majestic palms crowned in the autumn with ripening fruit. Dates are Iraq's second most valuable export, after petroleum.

In the south, where the Tigris and Euphrates meet with the Shatt al Arab, the land becomes a vast swamp with two marshy lakes, the Hawr al Hammar and the Hawr as

Saniyah. Several natural channels link the isolated settlements of Iraq's *madan* (marsh Arabs). The madan live in reed houses built on artificially made islands of mud and reeds. They paddle from one island to another in canoes.

The mountains

The mountains of northeastern Iraq form part of the range called Zagros in Iran and Taurus in Turkey. Some Iraqi peaks in this range rise to more than 11,000 feet (3,350 meters) above sea level.

The region is generally cold and wet. In winter, many of the mountain valleys are cut off from each other by heavy snow. In the milder months, spectacular waterfalls, forests

Along the Shatt al Arab waterways, *left,* date palms grow in large groves. The date palm is one of the oldest crop plants. According to Muslim legend, it represents the tree of life. A date palm can grow as tall as 100 feet (30 meters).

of walnut trees, and refreshing cold springs attract thousands of visitors.

The foothills and valley region are also called *Kurdistan* because the area is home to the Kurds. The Kurds are a rugged group of people who farm the land and herd sheep and goats for a living.

The desert

The desert covers southern and western Iraq. It also extends into Jordan, Kuwait, Saudi Arabia, and Syria. Most of this region of limestone hills and sand dunes is part of the Syrian Desert. Much of the land is almost completely uninhabited. Occasionally, villages and small towns can be seen near the old caravan trails. In recent years, highways have replaced the camel trails across the desert.

Tribes of Bedouin herdsmen still drive their sheep through the desert. Some tribes have abandoned the nomadic way of life to settle in agricultural communities.

Naour (water wheels) on the Euphrates provide necessary irrigation for Iraqi agriculture. These water systems date from the earliest civilizations. Modern concrete canals now supplement the water wheels, enabling farmers to extend their crops into drier areas.

The northern areas of Iraq, *far left,* are also known as Kurdistan. They make up the foothills and valleys of the Zagros Mountains. These forested valleys and cool uplands are quite different from the sands and stony plains of the Syrian Desert.

People

A variety of ethnic and religious groups live in Iraq. About 80 per cent of the population are Arabs. Kurds make up another 15 per cent. The remainder include Armenians, Assyrians, Sabians, Turkomans, and Yazidis. Almost half of Iraq's population is under 15 years of age. Although Arabic is the official language, many people speak Kurdish or Turkish. Quite a few are fluent in English, the international language of oil and commerce.

Religious life

About 95 per cent of Iraqis are Muslims. The majority belong to the Shiah branch of Islam. Most of the Shiites live in the central and southeastern regions of the country. The members of the other principal Muslim sect, the Sunnis, live mainly in the north.

One of the holiest Shiite shrines stands in the city of Karbala, some 60 miles (100 kilometers) south of Baghdad. For centuries, pilgrims from all over the world have come there to visit the shrine of Imam Husain, the Shiite saint who lived in the 600's. Women must wear the veil when crossing the square in front of the mosque and when walking through the town bazaar. In nearby An Najaf, 60 miles (100 kilometers) to the south, theology is taught in Shiite institutions of learning.

For decades, the Shiites have felt excluded from positions of national power and prestige. Their angry feelings began in the 1930's. During the Iran-Iraq war of the 1980's, Iranians hoped that a Shiite rebellion within Iraq would weaken their enemy. They were surprised to find that the Iraqi Shiites supported their government and the war effort. The patriotism of Shiite Iraqis led Western powers to look with favor upon Iraq. Westerners began to see Iraq as a barrier to the spread of Iran's militant Shiism.

The new and the old

Theological differences are not the only conflicts that split Iraq today. In parts of Baghdad, modern ways often come up against age-old traditions.

Baghdad's suq, or bazaar, is one of the few places in the city that has escaped modernization. For centuries, rich and poor alike have shopped the covered market for everything from spices to the wares of the goldsmiths and coppersmiths.

The differences between the new and the old can be seen in the position of women in Iraq today. It is not unusual to see a modestly veiled Iraqi mother with a daughter who wears Western fashions and cosmetics. The Baath Party guarantees equality for women, who now make up 25 per cent of the country's work force.

In contrast to the bazaar, modern luxury hotels line the banks of the Tigris River in Baghdad. These hotels draw the country's elite, who are attracted to the Western life style.

The discontent of the Kurds is another source of tension in Iraq. The Kurds would like to have their own separate Kurdish state. They regularly stage demonstrations to express their wish for political independence. In 1970, the Baath Party offered the Kurds radio and television programs in their own language, as well as Kurdish instruction in many schools. But for many Kurds, that was

A street vendor outside one of Baghdad's mosques, *far left,* wears the traditional headdress and baggy trousers of the Kurdish people. Iraq has the largest population of Kurds in the Middle East. They account for about 15 per cent of Iraq's total population. Most Kurds live in the foothills and valleys of the mountains in northeastern Iraq. This region is called Kurdistan.

The al Abbas Mosque, *left,* in Karbala, south of Baghdad, is one of the most important Shiite Muslim shrines. It contains the tomb of Imam Husain, a Shiite saint. Many Muslims visit Karbala before making a pilgrimage to Mecca. Shiites make up nearly two-thirds of Iraq's Muslim population. In contrast to the more orthodox Sunni Muslims, the Shiites worship a large number of saints.

A cafe in central Iraq, *left,* is a popular gathering place for villagers. Most Arab men and women wear long cotton gowns that reach to their ankles. Men cover their heads with a square cloth folded in half and held in place with a rope band.

not enough. They took sides with Iran against the Iraqi army in the Iran-Iraq war. After the war ended, the Kurds suffered harsh treatment at the hands of the Iraqi army for their actions.

The Persian Gulf War of 1991 triggered another Kurdish uprising in northern Iraq, but the Iraqi army swiftly put down the rebellion. More than a million Kurds then fled to the mountains, where thousands of them died of disease, exposure, or hunger.

A small group of people who speak Turkish live in the northern part of Iraq. But the few thousand that remain are becoming assimilated into the rest of Iraq's population. The Turkish-Iraqi border region is also home to the descendants of the ancient civilization of Assyrians and about 36,000 Yazidis, who are of Kurdish origin.

According to custom, an Iraqi woman wears a long, concealing cloak when she leaves home. She ties a scarf around her head and may cover her face with a veil.

Economy

Iraq's economy is built around oil and agriculture. Of these two sources of income, oil is by far the most important.

In the 1970's, the worldwide demand for oil was greater than the supply, and prices for oil soared. During this period, Iraq enjoyed a great boost in its economy. The government used some of these oil revenues to modernize Iraq's agriculture industry and make the country less dependent on imported food.

Oil and industry

Iraq is a member of the Organization of Petroleum Exporting Countries (OPEC). Before OPEC was founded in 1960, the petroleum industry in Iraq and other Middle Eastern countries was controlled by oil companies in the United States and Europe.

Countries like Iraq were paid royalties on the oil they produced. These royalties were based on the *posted price* set by the oil companies for crude oil. OPEC gave oil-producing nations more control over oil pricing and production.

In the 1970's, worldwide demand for oil was greater than the supply that could be produced by non-OPEC countries. OPEC nations, like Iraq, then dramatically raised their prices for crude oil.

This price increase created an economic boom in OPEC nations. In Iraq, in addition to modernizing agriculture, the government used oil profits to finance a rapid expansion of the country's steel, plastics, cement, and fertilizer industries. By the end of the 1970's, oil revenues had also established Iraq's textile and electronics industries and supported many public works projects, including hospitals, schools, roads, and airports.

The outbreak of the war against Iran in 1980 ended Iraq's oil boom, along with its dream of quick and easy profits. Just weeks after the war began, Iranian commandos destroyed Iraq's oil-loading facilities along the shores of the Persian Gulf.

Before the Persian Gulf War, large construction sites were changing the face of Baghdad. Iraq's government was also building new highways and improving the country's system of communication. Since the end of the war, most construction in Iraq has focused on rebuilding damaged buildings, bridges, and industries as well as restoring basic services such as electricity and water.

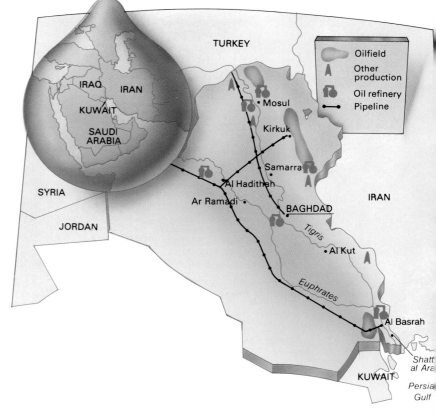

Iraq is one of the world's major oil-producing countries. The country's first oil field, near Kirkuk, began producing in 1927. Most of Iraq's oil reserves are in the oil fields near Al Basrah. The Iraqi government controls the country's oil industry.

Farmers grow cucumbers under plastic covers, *above,* near Samarra, along the Tigris River north of Baghdad. In the future, Iraq hopes to grow enough food to feed its people.

The nation's oil exports slowed to a trickle. Iraq's industries also suffered, and plans for building new cities and factories had to be postponed. Iraq's dependence on profits from crude oil led to severe financial losses during the war against Iran, when normal trade routes were interrupted.

Iraq's invasion of Kuwait in August 1990 caused another severe blow to the country's economy. Four days after the invasion, the United Nations Security Council imposed an embargo that prohibited all trade with Iraq except for medical supplies. Since nearly all of Iraq's major trading partners supported the embargo, Iraq's foreign trade all but ended.

After the Persian Gulf War ended, the UN continued the embargo to pressure Iraq into carrying out the terms of the formal cease-fire agreement. Under these terms, Iraq was to destroy all of its biological and chemical weapons and repay Kuwait for war damages.

Agriculture and manufacturing

About one-third of Iraq's people make their living from farming and herding. Iraqi farmers grow barley, dates, grapes, melons, olives, oranges, rice, tomatoes, and wheat. Herders raise camels, goats, horses, and water buffaloes.

Only about 9 per cent of Iraq's workers make their living in a manufacturing industry. Most Iraqi factories process farm crops and make such products as beverages, flour, leather goods, soap, and vegetable oil.

Traditional handcrafted objects, like those created by this carpenter in Souk-al Khadimiyah, are still an important part of Iraq's economy.

Mesopotamia

The valleys of the Tigris and Euphrates rivers in present-day southern Iraq were home to the world's earliest civilizations. They were the center of an ancient region called Mesopotamia—a Greek word meaning *between the rivers*. Mesopotamia included the area that is now eastern Syria, southeastern Turkey, and most of Iraq.

In spring, water from melting snow on Iran's high plateau causes the Tigris and Euphrates rivers to overflow their banks. Through the centuries, the regular flooding of the valley between the rivers created vast marshlands. Sometime before 3500 B.C., a group of people settled in these marshlands and developed the first civilization. This area came to be called Sumer.

The marshes where the Sumerians first settled are permanent swamplands. The main vegetation is a reed called *qasab*.

The Sumerians built irrigation ditches and canals that allowed water to be diverted to areas away from the rivers. Their irrigation system gave the Sumerians rich harvests of barley, wheat, dates, and vegetables. It also saved labor, which meant that people had time and energy for other activities.

The Sumerians' most important achievement was inventing the world's first writing system. From this first system of word pictures, the Sumerians developed a script called *cuneiform*. The Sumerians made cuneiform characters by pressing a tool with a wedge-shaped tip into wet clay tablets and then drying the tablets in the sun. Hundreds of thousands of these tablets have been found by archaeologists. They indicate that these ancient people had knowledge of mathematics, astronomy, and medicine.

By about 3500 B.C., the Sumerians had established several cities. These cities eventually grew into independent city-states. The more powerful city-states conquered their neighbors and set up independent kingdoms.

Wars between the city-states brought the downfall of Sumer during the 2300's B.C. The Akkadians, a Semitic people from the west, conquered Sumer and, along with other groups, ruled Mesopotamia between 2300 and 539 B.C. In 539 B.C., Meso-

A reconstruction of a Sumerian temple in the ancient city of Ur shows how these ancient people built temples called *ziggurats* in stepped formation.

potamia became part of the Persian Empire. Alexander the Great conquered the Persians in 331 B.C.

Around A.D. 750, Islam's Abbasid dynasty moved the Islamic capital from Damascus to Baghdad, on the Tigris River. Then, in 1258, the Mongols crushed the Abbasids and almost destroyed Baghdad. During this period, the marshes of Lower Mesopotamia became a stronghold of the Zanj, a people once used as slaves to drain the marshes around Al Basrah. In 1534, the Ottoman Turks seized the region.

Mesopotamia remained part of the Ottoman Empire until World War I (1914–1918), when the British took control. In 1921, most of Mesopotamia became part of Iraq.

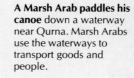

A Marsh Arab paddles his canoe down a waterway near Qurna. Marsh Arabs use the waterways to transport goods and people.

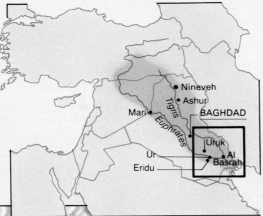

Mesopotamia, *right,* was the birthplace of civilization. The Sumerians first settled in the marshlands of Lower Mesopotamia, where they built reed houses and canoes.

The marshes of southern Iraq, *left,* were created by the seasonal flooding of two historic rivers, the Tigris and Euphrates.

Marsh Arabs, *right,* still inhabit the region once known as Lower Mesopotamia. They live in reed houses and have their own unique culture.

Reeds called *qasab* are the basic materials for most products used by the Marsh Arabs. These reeds, the main marsh vegetation, grow up to 25 feet (7.6 meters) tall.

The Kurds

The Kurds, a tribal society of farmers and herders, live in the mountainous regions of Southwest Asia. Their homeland, called Kurdistan, is officially only a small province in Iran. Yet the name *Kurdistan* is generally given to the area stretching from the southern border of Azerbaijan, through the Pontic and Taurus ranges in eastern Turkey, across northeast Syria and Iraq to the Zagros Mountains in northwestern Iran.

The number of Kurds living within this large area is generally estimated at about 20 million. In Iraq, Kurds make up about 18 per cent of the total population. In Turkey, they are the largest minority. The major Kurdish cities are Mahabad, Sanandaj, and Kermanshah in Iran; Irbil, Kirkuk, and As Sulamaniyah in Iraq; and Diyarbakir and Van in Turkey.

The majority of Kurds are Sunni Muslims. They speak Kurdish, an Indo-European language closely related to Persian.

History of the Kurds

The origins of the Kurds are uncertain. Historians believe that the mountains above Mesopotamia were settled around the 600's B.C. by a tribal people who fought—and sometimes defeated—the people of the plains. These tribal people were probably the early ancestors of the Kurds.

In the 1100's, the Seljuk Sultan Sandjar created a large province for these people and called it Kurdistan. Until the 1500's, the Kurdish tribes were under the control of the various powers who had conquered that part of Southwest Asia. In 1514, the Medes defeated the Assyrian Empire, and Kurdistan became a separate country between the Turkish and Persian empires.

But to this day, the Kurds have never had their own government. And because they want to be politically independent, they have often been in conflict with the governments under which they live.

Way of life

Most Kurds are farmers and herders. The farmers grow cotton, tobacco, and sugar beets. The herders raise sheep and goats.

Kurdish farming communities are led by an *agha,* who owns the village and the farmland. The farmers grow crops as their ancestors did, using wooden plows drawn by oxen. They live in simple houses made of mud bricks. The interiors are brightened by the handwoven rugs for which the Kurds are famous. Many Kurds still wear their colorful, embroidered traditional dress.

Kurdish women enjoy much more freedom than women in most Islamic societies. They do not wear the traditional veil, and they often hold important positions within their communities.

Iraqi Kurds, like this man, *far right,* carrying loaves of flat bread, make up about 18 per cent of the country's population. Most Kurds are Sunni Muslims.

Pastures in the mountains and valleys of Southwest Asia provide grazing land for the Kurdish herders' flocks of sheep and goats.

The Kurdish Struggle

Since the beginning of the 1900's, many Kurds have left their rural life in the mountains for jobs in the cities. Kurds have settled in the cities of Kermanshah in Iran, Kirkuk and Irbil in Iraq, and Diyarbakir in Turkey. There, they have continued the struggle for Kurdish political independence.

The struggle has its roots deep in history. Since the 1920's, there have been several major conflicts between the Kurds and the governments of the countries in which they live.

After World War I, the Turkish-Allied Treaty of Sèvres (1920) promised an independent Kurdistan. However, Kemal Atatürk, leader of the Turkish Nationalists, rejected the Treaty of Sèvres. He signed a new peace treaty with the Allies, the Treaty of Lausanne, which made no mention of the Kurds.

Since then, the Kurds have been in almost constant revolt. Kurdish uprisings in Iran followed the revolution of 1979, and fighting broke out in Turkey. Most recently, the struggle has been

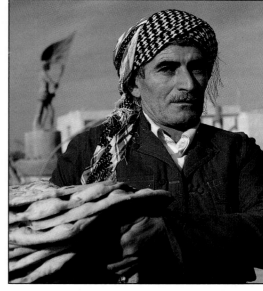

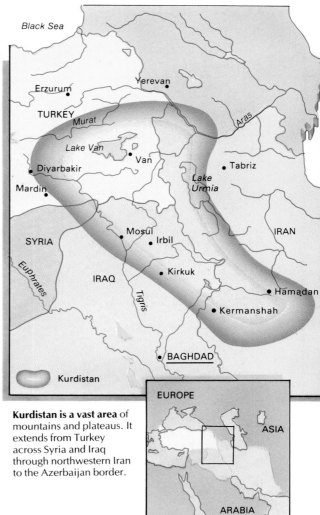

centered in Iraq. Under the leadership of Mustafa Barzani, the Kurds rose up in force in 1961. A truce was reached in 1970, and Iraq promised independence for the Kurds. But the agreement ended in a dispute over oil rights, since the Kurdish homeland has some of Iraq's richest oil reserves.

After the Iraq-Iran war ended in 1988, the Kurds suffered harsh treatment from Iraqi troops for their support of Iran during the conflict. After the Persian Gulf War, the Kurds once again rebelled against Iraq, but the revolt was brutally crushed by Iraqi president Saddam Hussein. After the revolt, 1 million Kurds fled to the mountains. There, thousands died of exposure, hunger, and disease.

Throughout their conflict with Iraq, Kurdish refugees have fled to Turkey for safety and protection, where they live in primitive camps, *left*. (A Kurdish independence fight within Turkey itself has killed more than 13,000 people since 1984.) In March 1995, Turkey invaded northern Iraq with the aim of destroying camps used by Kurdish guerrillas who had committed acts of violence within Turkey.

Kurdistan is a vast area of mountains and plateaus. It extends from Turkey across Syria and Iraq through northwestern Iran to the Azerbaijan border.

683

Ireland

Ireland, a small, independent country in northwestern Europe, covers about five-sixths of the island of Ireland. Dublin is its capital and largest city. The northeastern corner of the island, called Northern Ireland, is part of the United Kingdom of Great Britain and Northern Ireland.

Ireland is sometimes called the Emerald Isle because of its beautiful green countryside. Gently rolling farmland covers most of central Ireland, and mountains rise on its coasts. The picturesque ports of Cork and Waterford lie on natural harbors on the south coast, and Dublin Bay, on the east, serves as a harbor for Dublin, Ireland's major port. Many inlets and bays cut deeply into the more rugged west coast; among them are the beautiful Galway Bay and the mouth of the River Shannon. As a result, no point in the country is more than 70 miles (110 kilometers) from the sea.

The Irish are known for their friendliness and hospitality, as well as their skill in storytelling and writing. Ireland has produced a remarkable number of famous writers, including William Butler Yeats, Oscar Wilde, George Bernard Shaw, James Joyce, and Samuel Beckett. Irish folk music, known around the world, ranges from lively jigs and reels to age-old songs of love and sorrow and the emigrant's longing for home.

Saint Patrick, the *patron*—or guardian—saint of Ireland, brought Christianity to the country in the A.D. 400's. Today, about 95 per cent of the Irish people are Roman Catholics. For hundreds of years, the Irish Catholics were persecuted by the Protestant British. In 1919, Catholic Ireland declared its independence from Great Britain. In 1921, a treaty was signed that created an independent Ireland with its present borders. Predominantly Protestant Northern Ireland remained part of Great Britain. Many people in Ireland and many Catholics in Northern Ireland want their states to reunite, but the majority of the people of Northern Ireland want their country to remain separate.

Emigration has been a major problem in Ireland. The potato famines of the 1840's forced hundreds of thousands of people to leave the country. And in the early 1900's, thousands of people emigrated because of the limited job opportunities in Ireland. Since the 1920's, the development of new industries has helped check emigration, and the population has been increasing since the 1960's. However, a rise in unemployment in the 1980's once again resulted in a rise in the numbers of Irish people leaving their homeland.

Ireland Today

Ireland is a republic with a president, a prime minister known as the *taoiseach,* and a parliament. The president, the official head of state, is elected by the people to a seven-year term and may not serve for more than two terms. The powers of the president are limited to appointing the taoiseach, approving laws, and calling Parliament into session.

The president appoints the taoiseach, the head of government, who is usually the leader of the majority party in the legislature's lower house. The taoiseach selects other members of Parliament to serve in the Cabinet.

Parliament, called the *Oireachtas* in the Irish language, consists of the president, the legislature's lower house *(Dáil Éireann),* and the Senate *(Seanad Éireann).* The Dáil, which has 166 members elected by the people, makes Ireland's laws. The Senate serves mainly in an advisory capacity. Some senators are elected and some are appointed.

Ireland has five main political parties: *Fianna Fáil* (Soldiers of Destiny); *Fine Gael* (Gaelic People); the Labour Party; the Democratic Left; and the Progressive Democratic Party. Everyone 18 years of age or older may vote. Foreign residents may vote in local elections but, with the exception of

Irish-language areas (Gaeltacht)

Although English is used everywhere in Ireland, Irish is the everyday language in several areas—collectively known as the *Gaeltacht.*

The library of Trinity College, Dublin, contains the illuminated *Book of Kells* (c. A.D. 800) made by Irish monks.

FACT BOX

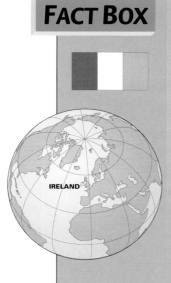

COUNTRY

Official name: Ireland
Capital: Dublin
Terrain: Mostly level to rolling interior plain surrounded by rugged hills and low mountains; sea cliffs on west coast
Area: 27,135 sq. mi. (70,280 km²)
Climate: Temperate maritime; modified by North Atlantic Current; mild winters, cool summers; consistently humid; overcast about half the time
Main rivers: Shannon, Liffey, Barrow, Boyne, Moy, Nore, Suir
Highest elevation: Carrauntoohil 3,414 ft. (1,041 m)
Lowest elevation: Atlantic Ocean, sea level

GOVERNMENT

Form of government: Republic
Head of state: President
Head of government: Prime minister
Administrative areas: 26 counties
Legislature: Oireachtas (Parliament) consisting of the Seanad Éireann (Senate) with 60 members serving five-year terms and the Dáil Éireann (House of Representatives) with 166 members serving five-year terms
Court system: Supreme Court
Armed forces: 11,500 troops

PEOPLE

Estimated 2002 population: 3,811,000
Population growth: 1.16%
Population density: 140 persons per sq. mi. (54 per km²)
Population distribution: 58% urban, 42% rural
Life expectancy in years: Male: 74 Female: 80
Doctors per 1,000 people: 2.3
Percentage of age-appropriate population enrolled in the following educational levels: Primary: 141* Secondary: 109* Further: 45
Languages spoken: English Irish (Gaelic)

British residents, may not vote in national elections.

On the local level, Ireland is divided into 26 counties and 5 county boroughs. Each county and county borough is governed by an elected council and a county manager appointed by the national government.

About 57 per cent of Ireland's people live in cities and large towns, although only Dublin and Cork have more than 100,000 inhabitants. The rest of the Irish people live in small rural towns and villages or in the countryside.

Ireland has two official languages—English and Irish, a Celtic language. All the people speak English. Although about 30 per cent consider themselves able to converse in Irish, only a small number use it as their everyday language. Since Ireland gained its independence in the early 1900's, many Irish people have sought to bring the Irish language into wider use. Today, the Irish government requires schools to teach Irish as well as English and uses

Ireland occupies five-sixths of the island of Ireland. The remainder of the island consists of Northern Ireland, which is part of the United Kingdom.

Religions:
Roman Catholic 95%
Church of Ireland
Presbyterian
Methodist

Enrollment ratios compare the number of students enrolled to the population which, by age, should be enrolled. A ratio higher than 100 indicates that students older or younger than the typical age range are also enrolled.

TECHNOLOGY

Radios per 1,000 people: 695

Televisions per 1,000 people: 399

Computers per 1,000 people: 359.1

ECONOMY

Currency: Euro

Gross national income (GNI) in 2000: $86.0 billion U.S.

Real annual growth rate (1999–2000): 11.5%

GNI per capita (2000): $22,660 U.S.

Balance of payments (2000): -$593 million U.S.

Goods exported: Machinery and equipment, computers, chemicals, pharmaceuticals, live animals, animal products

Goods imported: Data processing equipment, other machinery and equipment, chemicals; petroleum and petroleum products, textiles, clothing

Trading partners: European Union, United States, Japan

both English and Irish for official government business.

The Roman Catholic Church, which has long played a major role in Irish life, influences Irish law and operates many of the country's primary and secondary schools. The local church serves as a meeting place where many people take part in social events.

For hundreds of years, nearly all the Irish people made their living from farming, and many still do. However, agriculture has declined in importance since the 1920's, and today Ireland's economy depends heavily on service industries and manufacturing.

Ireland's most famous products include *stout,* a type of beer; delicate cut glass; high-quality linen, and fine woolen clothing. And the potato, which has long been important to Ireland's economy, is one of the country's major agricultural products.

History

The first people to live in Ireland probably came from the European mainland about 6000 B.C. About 400 B.C., invading Celtic tribes gained control of the island and divided it into small kingdoms.

In A.D. 432, Saint Patrick converted Ireland to Christianity. Saint Patrick also introduced the Roman alphabet and Latin literature into Ireland, and monasteries flourished there after his death. His feast day, March 17, is a national holiday in Ireland.

Around 795, the Vikings began a series of raids on Ireland, robbing and destroying the monasteries. The Irish could do little to defend themselves against the well-armed invaders until 1014, when the Irish king, Brian Boru, defeated the Vikings at Clontarf (now part of Dublin).

Norman nobles began seizing Irish lands in the late 1100's. By the 1300's, Normans controlled nearly all of Ireland.

English conquest of Ireland

In 1541, Henry VIII of England, in an attempt to gain control of Ireland, forced the Irish

c. 6000 B.C. First people settle Ireland from Europe.
c. 400 B.C. Celtic tribes invade Ireland.
A.D. 432 Saint Patrick converts the Irish to Christianity.
c. 795 Vikings begin a series of raids on Ireland.
1014 Vikings are defeated by Brian Boru.
1171 Normans recognize Henry II of England as lord of Ireland.

1541 Henry VIII of England is declared king of Ireland.

1690 James II and Irish forces are defeated at the Battle of the Boyne.

1801 Ireland becomes part of the United Kingdom of Great Britain and Ireland.
1845–1847 A potato famine kills about 750,000 people.

1916 The Easter Rebellion against British rule breaks out in Dublin but is put down within a week.
1919 Irish republicans establish a parliament in Dublin and declare Ireland an independent republic.
1920 Northern counties of Ireland become the state of Northern Ireland.
1921 Southern Ireland becomes a dominion of the United Kingdom called the Irish Free State.
1949 The Irish Free State cuts ties with the United Kingdom and becomes a republic, Ireland.
late 1960's Violence again breaks out between Protestants and Catholics in Northern Ireland.
1973 Ireland joins the European Community, now known as the European Union.
1985 Ireland is given an advisory role in Northern Ireland's government.
1990 Mary Robinson became Ireland's first female president.
1999 Ireland began taking part in new governing bodies created by the Northern Ireland peace agreement.

After Saint Patrick preached Christianity to the Celts in Ireland, the Roman Catholic monasteries in Ireland were a stronghold of Western civilization and art during Europe's Dark Ages. This Celtic cross dates from the 900's.

Jonathan Swift (1667–1745)

Daniel O'Connell, *far left,* (1775–1847)

William Butler Yeats, *left,* (1865–1939)

Parliament to declare him king of Ireland and tried to force Protestantism on the Irish. His daughter gave counties in Ireland to English settlers, and Elizabeth I outlawed Catholic religious services. This religious persecution continued through the 1600's.

James II, a Catholic king, attempted to restore the rights of the Irish Catholics, but he was forced from the English throne in 1688. When he tried to regain power with an army raised in Ireland, Irish Protestants helped the English defeat him in 1690.

In 1801, the Act of Union made Ireland part of the United Kingdom of Great Britain and Ireland. The Irish Parliament was dissolved, but the Irish eventually won the right to send representatives to the British Parliament.

Irish fishermen from the Aran Islands carry a *curragh,* a boat whose design dates back to prehistoric times, *left.* Early Irish settlers who moved to the island around 6000 B.C. lived by hunting and fishing.

Riflemen of the Irish Citizen Army, *below,* shoot from the roof of Liberty Hall, Dublin, at the time of the Easter Rebellion of 1916. Although British troops crushed the rebellion, it inspired the Irish people to win a "Free State" five years later.

Additional misfortunes afflicted the Irish people in 1845, when a plant disease wiped out the potato crop. From 1845 to 1847, about 750,000 people died of starvation or disease, and hundreds of thousands of others left the country.

Irish rebellion

During the late 1800's, some Irish people began to demand *home rule,* under which Ireland, as part of the United Kingdom, would have had its own parliament for domestic affairs. The British Parliament twice rejected home rule bills.

Republicans—Irish who wanted a completely independent Irish republic—rebelled against the British in Dublin on Easter Monday of 1916. The Easter Rebellion was easily put down, but in 1918 the republicans won most of Ireland's seats in the British Parliament. Instead of going to London to take their seats, they formed the *Dáil Éireann* (House of Representatives) in Dublin, and in 1919 they declared all Ireland an independent republic.

As a result, fighting broke out between British forces and the Irish Republican Army (IRA), as the rebels were called. In 1920, the British Parliament passed an act that divided Ireland into two separate countries—one consisting of six northern counties and the other of the rest of Ireland. The northern counties accepted the act and formed the state of Northern Ireland. The southern counties rejected the act and fought on. Finally, in 1921, the United Kingdom signed a treaty making southern Ireland a dominion (self-governing country) of the British Commonwealth called the Irish Free State. In 1949, all ties with the United Kingdom were finally cut, and Ireland became an independent republic.

Bitterness between Catholics and Protestants in Northern Ireland erupted into violence in the late 1960's. In 1994, paramilitaries on both sides declared a cease-fire. However, the IRA broke the 17-month cease-fire when it resumed terrorist bombings in February 1996. In 1998, peace talks on Northern Ireland concluded in an agreement that promised an end to the conflict in the troubled region. In 1999, the United Kingdom ended direct rule of Northern Ireland. The Republic of Ireland, in turn, gave up its claim to Northern Ireland.

The Irish Character

The Irish are generally considered to be warm, friendly, talkative people. Legend connects their "gift of gab" with the Blarney Stone—a block of limestone in Blarney Castle, near Cork. According to the legend, anyone who kisses the stone receives the gift of expressive, convincing speech. Today, the word *blarney* has come to mean flattering or coaxing talk.

Literature and music

The traditional Irish way with words also extends to writing, and Ireland has produced many great writers over the centuries. In the late 1800's, a number of authors set out to create a body of literature in

Many Irish people are fond of horses, and horse racing is a popular sport in Ireland. Steeplechasing, in which horses leap over ditches, hedges, and other obstacles, is a favorite track event.

the English language that would express the Irish experience and tradition.

During this period of Irish literary revival, the poet William Butler Yeats and the playwright Lady Gregory founded Dublin's Abbey Theatre. The theater produced dramas by Sean O'Casey, John Synge, Yeats, and other gifted Irish playwrights. Some of the plays dealt with Irish peasant life, while others, such as O'Casey's *The Shadow of a Gunman* (1923) and *Juno and the Paycock* (1924), reflected the political upheaval raging in Ireland during the 1920's.

Music is also a vital part of Irish life, especially the folk music played on the harp and the bagpipes. Irish folk songs can as easily tell the tale of a struggling farmer or a lost love as they can relate a lot of comical "blarney" about a fast-talking Irishman.

Everyday life

While most of the people in Irish cities and towns live in houses, apartment living is on the rise. Many apartments are located in buildings above grocery stores and other shops.

At one time, large numbers of young people lived with their parents and remained single until they were well over 30, because farmland and jobs were scarce, and few young people could afford to marry and raise families. Today, with more job opportunities opening up in the cities and towns, many young people leave home and marry earlier.

Modern houses have replaced most of the traditional thatch-roofed cottages that once dotted Ireland's countryside, but people in rural areas live much as their ancestors did. For instance, peat, which is used as fuel in Ireland, is still cut by hand and transported in baskets carried by donkeys.

In their leisure time, many Irish people enjoy sports, and horse racing is one of their favorites. Races are held about 170 days a year on some 30 race tracks around the country; the most famous events include the Irish Derby, held at Kildare in late June, and the Irish Grand National, held near Dublin on the Monday after Easter. Horse shows are also well attended, and thousands of visitors enjoy the Royal Dublin Society Horse Show held in Dublin each August.

Ireland's favorite team sports are soccer; *Gaelic football,* which resembles soccer; and *hurling,* which is similar to field hockey. Other popular team sports include *camogie,* which is also similar to field hockey and is played by women; cricket; and Rugby football. Many Irish people also enjoy boxing, fishing, and golf.

A strolling musician, *above left,* entertains at the Whitegate Pony Races in County Clare. Folk songs and traditional music are popular among the young Irish people also.

A young man in Connemara, County Galway, *above,* loads peat on his donkey's back just as his ancestors did. Peat, which consists of decayed plant material—and is an early stage in the formation of coal—is used as a household fuel.

A grocer in Athlone, the biggest town in County Westmeath, proudly displays his wares, *left.* Many shopkeepers in towns and cities live in apartments above their shops.

Israel

Israel is a small country on the eastern shores of the Mediterranean Sea that was created in 1948 as a homeland for Jewish people. Israel is a young nation, but its roots actually go back thousands of years. Jews all over the world consider Israel their spiritual homeland.

The Hebrew people settled in the area, which was then called Canaan, 4,000 years ago. Part of the Hebrews' land came to be named Judah, and the people came to be called Jews. Most Jews fled their homeland after A.D. 135, when the Romans put down Jewish revolts against their rule. The Jews were unable to recreate their nation until 1948, when they began to build a modern democratic state with a developed economy.

Modern Israel was created as a homeland for the Jewish people after World War II, so most of its people are Jews. However, they belong to a variety of ethnic groups, and Arabs make up an important ethnic minority. The land of Israel—and especially the city of Jerusalem—are sacred places to three major world religions: Judaism, Christianity, and Islam.

The creation of Israel caused problems in the Middle East. Most Arabs opposed its founding, and the result has been a number of wars between Israel and Arab nations—in which Israel gained territory—as well as much terrorist violence against Israel. In addition, Israel has problems with the Arab minorities in its own population. However, new hope for an end to the fighting dawned in 1993, when Israel and the Palestine Liberation Organization (PLO) signed a framework peace accord.

Israel has a pleasant climate, with hot, dry summers and cool, mild winters. The country has four major land regions: the coastal plain, the Judeo-Galilean Highlands, the Rift Valley, and the Negev Desert.

The narrow coastal plain is home to most of the nation's people, industry, and farmland. Here are two of Israel's three largest cities, Haifa and Tel Aviv, and the fertile plains of Esdraelon and Sharon.

East of the coastal plain are highlands and hills that run from Galilee in the north to the Negev Desert in the south. Most of Israel's Arabs live in Galilee. Both Nazareth, the largest Arab center, and Jerusalem, the country's largest city, are in the highlands.

The wedge-shaped Negev is an arid area of flatlands and mountains. But ingenious irrigation is turning sections of this desert into fertile farmland.

East of the highlands and desert lies the Rift Valley, which extends into Africa. The edges of the valley are steep, but the floor is flat and low. In the northern Rift Valley, the River Jordan flows through the Sea of Galilee to the Dead Sea. The shore of the Dead Sea is the lowest land area on earth.

Israel Today

Although Israel is a small country with few resources, its people have created a modern, democratic, economically developed nation. Israel has no written constitution. The government follows "basic laws" passed by its parliament, the Knesset, which makes laws and helps form national policy. The 120 members of the Knesset are elected for up to four years.

All Israeli citizens 18 years old or older may vote, but Israeli voters cast ballots for party lists, rather than for individual candidates. A party list includes all the candidates —from 1 to 120—of a particular political party. Israel has many different parties reflecting many different views, but two parties dominate national elections—the Labor Party and the Likud bloc.

The socialist Labor Party supports government control of the economy with some free enterprise. It also favors negotiating and compromising with neighboring Arab states.

The Likud bloc, on the other hand, supports limited government control of the economy and takes a hard line toward the Arab states. The bloc is actually an alliance of a number of smaller parties.

The outcome of an Israeli election is determined by the percentage of the vote given to each party list. If, for example, a particular party list received one-third of the vote, that party would control one-third of the Knesset, or 40 seats.

Usually, the leader of the party that controls the most seats becomes prime minister. The prime minister forms and heads the Cabinet, whose members head government departments and serve as Israel's top policy-making body. Israel also has a president elected by the Knesset, but the office is largely ceremonial.

The prime minister must have the support of a majority of the Knesset to stay in power. If the Labor Party or Likud bloc has too few seats to form a majority, it usually seeks support from the minor religious or special-interest parties in Israel. In this way, these small parties have considerable power.

Education has a high priority in Israel. One of the first laws passed in the nation established free education and required children between the ages of 5 and 15 to attend school.

The people of Israel enjoy a fairly high standard of living and a low level of unemployment. Israel began as a poor country, but its vigorous people have established industries, drained swamps, and irrigated the desert.

Today, Israel has a largely service economy. About 65 per cent of Israeli workers are employed in service industries, such as

FACT BOX

COUNTRY

Official name: Medinat Yisra'el (State of Israel)
Capital: Jerusalem
Terrain: Negev desert in the south; low coastal plain; central mountains; Jordan Rift Valley
Area: 8,019 sq. mi. (20,770 km²)

Climate: Temperate; hot and dry in southern and eastern desert areas
Main river: Jordan
Highest elevation: Mount Meron, 3,963 ft. (1,208 m)
Lowest elevation: Dead Sea, 1,310 ft. (399 m) below sea level

GOVERNMENT

Form of government: Parliamentary democracy
Head of state: President
Head of government: Prime minister
Administrative areas: 6 mehozot (districts)

Legislature: Knesset (Parliament) with 120 members serving four-year terms
Court system: Supreme Court
Armed forces: 173,500 troops

PEOPLE

Estimated 2002 population: 6,425,000
Population growth: 1.67%
Population density: 790 persons per sq. mi. (305 per km²)
Population distribution: 90% urban, 10% rural
Life expectancy in years: Male: 77 Female: 81
Doctors per 1,000 people: 3.9
Percentage of age-appropriate population enrolled in the following educational levels: Primary: 107* Secondary: 89 Further: 49

ISRAEL

community work, trade, or tourism. Many of these workers are employed by the government or work in businesses owned by the government.

About 25 per cent of Israel's business firms are government owned. Another 25 per cent are owned by the *Histadrut* (General Federation of Labor), a powerful organization of trade unions. Half of the nation's businesses are privately owned.

About 22 per cent of Israeli workers are employed in manufacturing. They make such goods as chemical products, electronic equipment, fertilizer, paper, plastics, processed foods, scientific and optical instruments, and textiles and clothing. The government-owned plants produce equipment needed by Israel's armed forces to maintain military readiness.

Israelis have transformed their dry, unproductive land into farmland through irrigation. Using modern methods and machines, a small number of farmers now produce enough food to meet domestic demands and pay for any necessary food imports. Fruits, cotton, eggs, grains, poultry, and vegetables are the main food products.

Despite its forbidding name, the Dead Sea is a source of mineral wealth. Table salt and *potash,* a mineral used for fertilizer, are extracted from the water.

Diamond cutters pursue their craft in a workshop in Israel, where cutting imported diamonds is a major industry. Other Israeli industries produce such high-technology goods as electronic equipment and scientific instruments.

Israel is a small Middle East country on the eastern shore of the Mediterranean Sea. Since it officially came into existence on May 14, 1948, it has fought several wars with its Arab neighbors.

Languages spoken:
 Hebrew (official)
 Arabic (official for Arab minority)
 English

Religions:
 Jewish 80%
 Muslim (mostly Sunni) 15%
 Christian 2%

Enrollment ratios compare the number of students enrolled to the population which, by age, should be enrolled. A ratio higher than 100 indicates that students older or younger than the typical age range are also enrolled.

TECHNOLOGY

Radios per 1,000 people: 526

Televisions per 1,000 people: 335

Computers per 1,000 people: 253.6

ECONOMY

Currency: new Israeli shekel

Gross national income (GNI) in 2000: $104.1 billion U.S.

Real annual growth rate (1999–2000): 6.0%

GNI per capita (2000): $16,710 U.S.

Balance of payments (2000): -$1,416 million U.S.

Goods exported: Machinery and equipment, software, cut diamonds, chemicals, textiles and apparel, agricultural products

Goods imported: Raw materials, military equipment, investment goods, rough diamonds, fuels, consumer goods

Trading partners: United States, United Kingdom, Benelux, Hong Kong, Germany

History

Twelve Hebrew tribes created the Kingdom of Israel about 3,000 years ago, but the kingdom later split in two. The Hebrews—who came to be called Jews—fell under the rule of a series of empires. Assyrians, Babylonians, Persians, and Greeks all conquered and ruled the Jews. The last of the conquerors were the Romans. In A.D. 135, Rome drove the Jews out of their capital of Jerusalem and renamed the area Palestine.

The move to establish the modern state of Israel began in the late 1800's, when Theodor Herzl, an Austrian, founded the Zionist movement. Zionists sought to establish a homeland for the world's Jews in Palestine.

In 1917, Great Britain issued the Balfour Declaration, stating its support for such a Jewish homeland. At the time, Britain was trying to win control of Palestine from the Ottoman Empire during World War I (1914–1918).

Palestine came under the administration of Great Britain after the war ended. Large numbers of Jewish immigrants fled to Palestine during the 1930's to escape Nazi persecution. Arabs living in the region became alarmed and rebelled against the British.

During World War II (1939–1945), millions of Jews were killed by the Nazis. After the war, demands for a Jewish state increased, and Great Britain turned the problem over to the United Nations (UN). The UN devised a plan to split Palestine into a Jewish nation and an Arab nation.

In 1948, the State of Israel was established under the leadership of David Ben-Gurion. But Arab nations refused to recognize Israel and attacked the new state. Israel quickly defeated the Arabs and gained control of about half the land planned for the new Arab state. About 150,000 Palestinian Arabs thus became part of Israel's population.

In 1967, Egypt sent troops into the Sinai and blocked an Israeli port. Israel launched an air strike and almost completely destroyed the air forces of Egypt and its allies, while its ground troops defeated their armies.

By the war's end, Israel had won control of

Jewish refugees from Europe flocked to Palestine after World War II, despite efforts by the British to limit Jewish immigration to the area. The newcomers helped found the State of Israel in 1948.

Israel's borders have varied since the country was established in 1948. When the UN plan to divide Palestine into Jewish and Arab states failed in 1948, Israel won a victory as well as additional territory in its war against surrounding Arab nations. In 1967, in only six days, Israeli forces took the Sinai Peninsula, Gaza Strip, Golan Heights, and West Bank. The 1979 peace treaty with Egypt resulted in staged withdrawal of Israeli troops from the Sinai. Israel holds the other territories and has established settlements in the West Bank.

1900–1700 B.C. Hebrews (Israelites) settle in Canan.
1600–1500 B.C. Some Hebrews move to Egypt.
1200's B.C. Moses leads Hebrews out of Egypt.
1000 B.C. David forms unified Kingdom of Israel.
900's B.C. King Solomon builds Temple.
922 B.C. Nation splits into states of Israel and Judah.
722 or 721 B.C. Assyrians conquer Israel.
587 or 586 B.C. Babylonians conquer and exile Jews. Solomon's Temple is destroyed.
167 B.C. Maccabeans lead Jewish revolt and establish kingdom of Judah.
63 B.C. Romans occupy Judah (Judea).
66 Romans put down Jewish revolt.
135 Romans drive Jews out of Jerusalem.
1800's Zionist movement seeks to make Palestine a Jewish state.
1917 Balfour Declaration states British support for Jewish homeland.
1920 Palestine comes under British administration.
1930's Jewish immigration sparks Arab violence.
1947 UN divides Palestine into Jewish state and Arab state.
1948 State of Israel created.
1948–49 Israel wins territory in first Arab-Israeli war.
1956 Israel, Great Britain, and France attack Egypt in second Arab-Israeli war.
1964 Palestine Liberation Organization (PLO) founded.
1967 Israel wins control of Sinai, Golan Heights, and West Bank in Six-Day War.
1973 Israel fights Egypt and Syria in Yom Kippur War.
1978 Israel and Egypt agree to the Camp David Accords.
1979 Israel and Egypt sign peace treaty.
1982 Israel attacks PLO camps in Lebanon.
1985 Israeli troops leave most of Lebanon.
1987 Intifada begins.
1990 Police kill 21 Palestinian demonstrators.
1991 Iraq launches missile attacks on Israel.
1993 Israel and PLO sign peace accord for interim Palestinian self-rule.
1995 Israeli Prime Minister Yitzhak Rabin assassinated.
2001 Ariel Sharon elected prime minister.

Theodor Herzl (1860–1904) founded the Zionist movement.

David Ben-Gurion (1886–1973) was Israel's first prime minister.

Golda Meir (1898–1978) led Israel through the fourth Arab-Israeli war.

The fourth Arab-Israeli war broke out in 1973. Egypt and Syria attacked Israel on Yom Kippur, the most sacred Jewish holy day. Israel pushed the Arab forces back. Tensions eased after the war, and in 1978 Egyptian President Anwar el-Sadat and Israeli Prime Minister Menachem Begin met with United States President Jimmy Carter. Their discussions resulted in the Camp David Accords, and a peace treaty was signed in 1979.

In 1993, Israeli Prime Minister Yitzhak Rabin and Yasir Arafat, chairman of the PLO, signed an accord in which Israel would cede administrative control of Palestinian land gained in the 1967 war to the Palestinians. In 1994, Israel and Jordan signed a nonaggression pact. But in 1995, Rabin was assassinated by an Israeli Jew who opposed the peace process. Following the assassination, Shimon Peres became prime minister. Benjamin Netanyahu, a critic of the Israel-PLO peace agreements, defeated Peres in 1996. Netanyahu's decision to resume construction of Israeli settlements in the West Bank was met with protests by Palestinians. In 1997, however, Israel completed an agreement with the PLO over the withdrawal of Israeli troops from most of the West Bank city of Hebron.

In 1998, Netanyahu, claiming that the PLO was not fulfilling its security commitments, suspended Israeli troop withdrawals. In 1999, Ehud Barak, leader of the Labor Party, was elected prime minister of Israel. Barak favored renewing the peace process with the Palestinians, however, the two sides were unable to agree on key issues. In an election in 2001, Ariel Sharon, the leader of the right-wing Likud Party, defeated Barak. Sharon formed a coalition government that included the Labor Party and several other parties.

Numerous attacks by Palestinian militias and suicide bombers took place throughout Israel, the West Bank, and the Gaza Strip, killing hundreds of Israelis. Israeli forces repeatedly bombed and invaded the West Bank and Gaza Strip, killing more than 1,800 Palestinians. In 2002, Israel reoccupied much of the West Bank.

In October 2002, the Labor Party withdrew from Sharon's coalition government. In the 2003 elections, Sharon's Likud Party won the largest number of seats. Sharon formed a new coalition government and remained as prime minister.

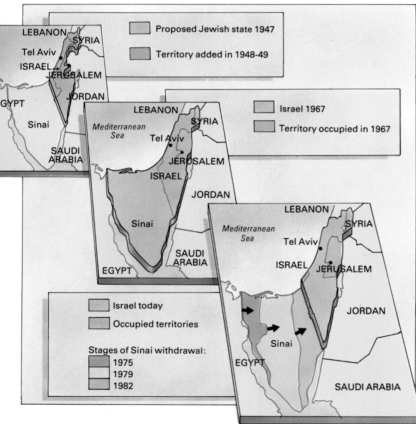

Egypt's Sinai Peninsula and Gaza Strip, Syria's Golan Heights, and Jordan's West Bank. Yet with this land came 1 million more Palestinian Arabs. The Palestine Liberation Organization (PLO) became especially active and launched attacks on Israelis.

People

When Israel was established in 1948, about 806,000 people lived there. Today, the population numbers nearly 6 million.

About 1.8 million Jews migrated to Israel in its first 40 years, including many who fled from persecution in their home countries. The Israeli government allows any Jew, with a few minor exceptions, to settle in Israel.

Heritage

Today, about 80 per cent of Israel's people are Jews, but while they share a common spiritual and historical heritage, they differ in their religious observance. About one-fifth of the Jewish population maintains strict observance of the principles of Judaism, and about half of the Jewish people observe some, but not all, of these principles. The rest of Israel's Jewish population is *secular* (nonreligious). These different groups disagree on the role Jewish religion should play in the government of Israel.

Because Israel's Jews came from many different countries, they belong to many different ethnic groups—each with its own cultural, political, and recent historical background. But Israel's Jewish population falls into two major cultural groups—the Ashkenazim and the Sephardim.

The Ashkenazim came to Israel from Europe or North America and are descended from members of the Jewish communities of central and eastern Europe. Many Ashkenazim speak Yiddish, a Germanic language. The Sephardim, on the other hand, came from countries in the Middle East and the Mediterranean.

A divided society

Today, most Israeli Jews are Sephardim, but in the early years of the nation, the Ashkenazim dominated. Many of these early Ashkenazim were skilled tradespeople and professionals who made a great contribution to Israel's development. As a result, Israel's political, educational, and economic systems are much like those of Western nations. The Sephardim have had to adapt to this "foreign" society.

Another group that has had a difficult time adapting to Israeli society is the Arab population. Arabs are by far the largest minority in Israel, making up nearly all the non-Jewish

population. Most Arabs are Palestinians whose families remained in Israel after 1948.

The Jews and Arabs of Israel are suspicious of each other—and sometimes openly hostile. Most live in separate areas, speak different languages, attend different schools, and follow different cultural traditions.

Israel has two official languages—Hebrew, the language spoken by most of the Jews, and Arabic, spoken mainly by the Arabs. The nation has two school systems—a Jewish system, in which instruction is given in Hebrew, and an Arab system, in which instruction is given in Arabic. The government recognizes and funds both systems. Although Israel's government guarantees religious freedom and allows members of all faiths to observe their Sabbaths and holy days, tension between the Jews and Arabs remains.

Urban life

About 90 per cent of Israel's people live in urban areas. About 25 per cent of the nation's population live in the country's

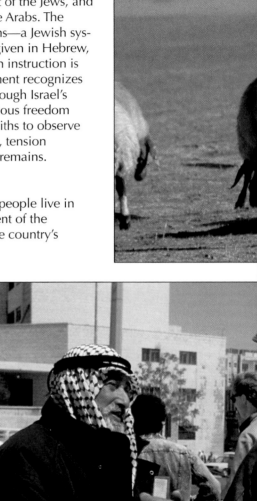

An Israeli soldier carries out a security check on an Arab resident of Jerusalem following rioting in the West Bank. Many Palestinian Arabs oppose Israeli rule.

Sephardic (eastern) Jews, like this Yemeni bride, follow customs different from those of *Ashkenazi* (European) Jews.

three largest cities—Jerusalem, Tel Aviv, and Haifa.

Many Israeli cities are built on the sites of ancient settlements where there are many historic buildings, but Israelis have also constructed large, modern sections with high-rise office buildings. Most urban Israelis live in modern apartments. They face the same problems as people in other rapidly growing urban areas, including traffic congestion, housing shortages, and pollution.

Although most Israelis wear Western-style clothing, their food and drink reflect their ethnic diversity. For example, traditional European Jewish dishes, such as chopped liver, chicken soup, and gefilte fish, are common. But so are traditional Middle Eastern foods such as *felafel*—small, deep-fried patties of ground chickpeas. However, all government buildings and most restaurants and hotels serve only *kosher* foods, which are prepared according to Jewish laws.

A Palestinian Arab tends her sheep in the Negev Desert. The Arabs live mainly in their own rural villages or in separate neighborhoods in Israeli cities.

Fishermen on the Sea of Galilee carry on a way of life that goes back to Biblical times. Most Israelis enjoy a relatively high standard of living, with income levels comparable to those in countries such as Spain and Greece.

Jews from all over the world have come to live in Israel. The Law of the Return, passed by the Knesset in 1950, allows any Jew, with minor exceptions, to settle in the country. A 1970 amendment to the law defined a Jew as "a person who was born of a Jewish mother or has become converted to Judaism and who is not a member of another religion." The Israeli government provides temporary housing and job training to immigrants.

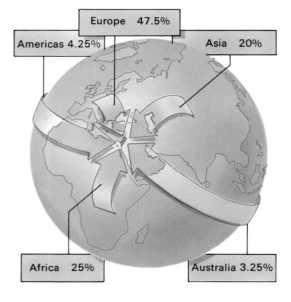

Europe 47.5%

Americas 4.25%

Asia 20%

Africa 25%

Australia 3.25%

Jerusalem

Jerusalem, Israel's capital city, is a holy city to Jews, Christians, and Muslims. This ancient city has been the site of much conflict through the ages.

When modern Israel was established in 1948, Tel Aviv was named its capital, and Jerusalem was to be an international city. But when Israel was attacked by neighboring Arab nations, Jerusalem became the scene of fierce fighting. By the end of 1948, Israeli soldiers held West Jerusalem, and Jordanian troops controlled East Jerusalem. The Israelis then declared West Jerusalem their capital city, but most countries that had diplomatic relations with Israel refused to recognize West Jerusalem as the nation's capital.

When war between Israel and Arab nations broke out again in 1967, Israel captured East Jerusalem and announced it would keep all of Jerusalem. In 1980, the Israeli government officially named the city of Jerusalem as its capital.

A sacred city

Israelis felt deeply that Jerusalem should be their capital because it is a holy city to Jews. Jerusalem was their political and religious center in Biblical times. About 1000 B.C., King David captured the town from a people called the Jebusites and made it the capital of the Kingdom of Israel. David's son, King Solomon, built a magnificent house of worship —the first Temple of the Jews—in the city.

Over hundreds of years, the Temple was captured, recaptured, destroyed, and rebuilt several times. Finally, to put down a Jewish revolt, the Romans burned the Temple in A.D. 70.

Today, the Western Wall, also called the Wailing Wall, is all that remains of the Jews' holy Temple. The wall, which is 160 feet (49 meters) long, was the western wall of the Temple courtyard and has long been a symbol of Jewish faith and unity. When Jerusalem was split after the war of 1948, many Jews were bitterly disappointed because the wall lay in Arab-controlled East Jerusalem.

Christians also regard Jerusalem as a holy city because Jesus Christ was crucified there. The Church of the Holy Sepulcher is believed to stand on the hill of Calvary, or

Jerusalem, as seen from the air, displays the power of religious faith. In the center stands the walled Old City, where monuments of three great world religions—Judaism, Christianity, and Islam—stand. About 75 per cent of Jerusalem's people are Jews, who live in the more modern neighborhoods of West Jerusalem. The population of East Jerusalem, where the holy sites are located, is mostly Arab.

Orthodox Jews walk through the narrow cobblestoned streets of the Old City, *bottom right.* Orthodox Jews believe in strict observance of the traditions and principles of Judaism. They sometimes oppose Jews who follow more secular ways.

Jews pray at the Western Wall, all that remained of the Temple of the Jews after the Romans burned it down in A.D. 70. The wall is also sometimes called the Wailing Wall, for the sorrowful prayers said there. According to legend, the wall itself weeps over the Temple's destruction.

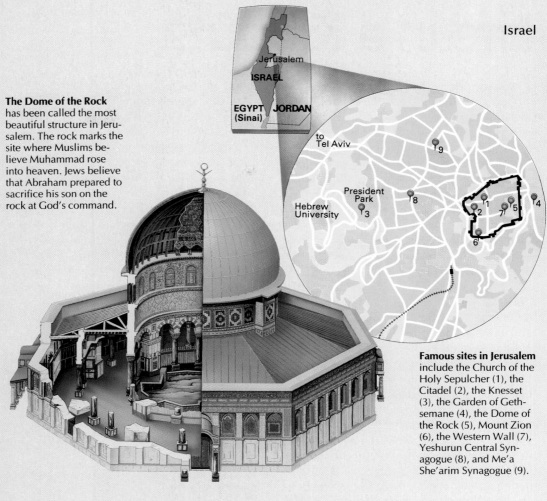

The Dome of the Rock has been called the most beautiful structure in Jerusalem. The rock marks the site where Muslims believe Muhammad rose into heaven. Jews believe that Abraham prepared to sacrifice his son on the rock at God's command.

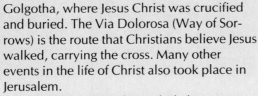

Famous sites in Jerusalem include the Church of the Holy Sepulcher (1), the Citadel (2), the Knesset (3), the Garden of Gethsemane (4), the Dome of the Rock (5), Mount Zion (6), the Western Wall (7), Yeshurun Central Synagogue (8), and Me'a She'arim Synagogue (9).

Golgotha, where Jesus Christ was crucified and buried. The Via Dolorosa (Way of Sorrows) is the route that Christians believe Jesus walked, carrying the cross. Many other events in the life of Christ also took place in Jerusalem.

To Muslims, Jerusalem is holy because they believe that Muhammad, the founder of their religion, rose to heaven from the city. A beautiful golden-domed monument called the Dome of the Rock marks the site. According to Muslim belief, Muhammad rose with the angel Gabriel and spoke to God, then returned to spread the religion of Islam. A 1994 peace pact with Israel gave Jordan a role in administering the Dome of the Rock. This was denounced by Palestinians, who felt it weakened their claim to part of the city.

Modern-day Jerusalem

Today, about three-fourths of the people of Jerusalem are Jews. They live in West Jerusalem, the newer section of the city, where some Christians and Muslims also live.

Almost all the people of East Jerusalem are Arabs. About four-fifths of them are Muslims, and most of the others are members of various eastern Christian churches. East Jerusalem includes the oldest district in all Jerusalem—the Old City.

The Old City lies on the site of ancient Jerusalem and is surrounded by stone walls almost 40 feet (12 meters) high and 2-1/2 miles (4 kilometers) long. Within these walls lie the Western Wall, the Church of the Holy Sepulcher, and the Dome of the Rock.

Jerusalem is a city of three Sabbaths—Friday (Muslim), Saturday (Jewish), and Sunday (Christian). Stores and businesses may be closed on any of these three days. After the Jewish Sabbath begins on Friday night, a large portion of West Jerusalem closes down, and there is no public transportation. Most Jews observe the Sabbath, but hundreds of others go to East Jerusalem where cafes and other places of entertainment are open.

Life in a Kibbutz

For many people throughout the world, the *kibbutz* symbolizes the nation of Israel. In Israel, a kibbutz is a community in which no one owns private property. All property belongs to the kibbutz, and the members work for the kibbutz. In return, the kibbutz meets all the needs of the members and their families. It provides food, housing, education, child care, and medical care.

The first kibbutz was founded in what was then Palestine in 1909. Two women and 10 men from Poland established a collective settlement called Deganya on the Sea of Galilee. Like other early kibbutzim, this settlement was an agricultural community. Today, more than 250 kibbutzim are found in Israel, and most own factories as well as farmland.

Kibbutzim range in size from about 50 to 1,500 members, but a typical kibbutz has 250 members. Israelis who wish to join a kibbutz usually work on the kibbutz for a year as a candidate for membership. Members of the kibbutz then vote on whether to accept the candidate.

All kibbutz members have an equal say in how the community is run. Many kibbutz members have regular daily jobs, while others are assigned to a variety of jobs by a work committee. Members receive no pay for their work but receive all the goods and services they need.

In some kibbutzim, children sleep in their parents' homes. In other kibbutzim, they sleep in children's houses. In either case, children spend most of the day with their peers. From kindergarten on, their education emphasizes the importance of cooperation. Children are assigned duties, and by high school spend one day each week working in the kibbutz.

Mothers of infants visit their children frequently during the day, and both parents join their children for a time after work. In this way, parents and children form close ties even when they do not live together.

Israelis who have been born and raised on kibbutzim are known for being high achievers. Many kibbutz members have served in the Knesset or the Cabinet, and others have distinguished themselves in the army. Some people believe these achievements are due to the superior education and health care received on kibbutzim. Also, kibbutz members value the group more than the individual, and learn to work well in many different group situations.

Only about 10 per cent of Israelis live in rural areas. Of these, more than half live in kibbutzim or in *moshavim* or *moshavim shitufi,* two other kinds of rural communities. Together, the kibbutzim and moshavim help produce almost all the food Israel needs.

In a moshav, while each family works its own land separately and lives in its own house, the entire moshav owns and shares the large pieces of farm equipment. The moshav purchases essential supplies, such as

Crates of peaches are packed for export at an Israeli farm. Most farms are organized as either moshavim or kibbutzim. On moshavim, the land is privately owned; on kibbutzim, the land belongs to the entire community.

seeds, and markets all the crops. Moshav members elect assemblies to supervise the moshav.

A moshav shitufi is like a moshav in some ways and a kibbutz in others. As in a moshav, members of a moshav shitufi have their own family households. As in a kibbutz, the members work collectively on the land, and profits are shared.

An aerial view of a kibbutz on the shores of the Sea of Galilee shows its green, well-tended fields. Although they had to struggle to feed their members in the early years, kibbutzim are now modern, well-equipped farms that help feed the entire nation.

Bringing in the harvest, a young Israeli woman hoists large containers that soon will be filled with produce. Because life in a kibbutz emphasizes equality, men and women are expected to share all kinds of work.

Volunteer workers take a break from their labor on a kibbutz. Young people from many countries take working vacations in such settlements in order to experience communal life.

Kibbutzim and moshavim make up most of Israel's farmland. Although early kibbutzim began as purely agricultural settlements, most now combine light industry and services with farming. They are collective communities, in which land, buildings, equipment, and money are owned by all the members of the kibbutz as a group. Committees are elected to run the kibbutz and decide how profits are to be distributed among members. Jewish pioneers from Poland set up the first kibbutz, called Deganya, near the Sea of Galilee in 1909.

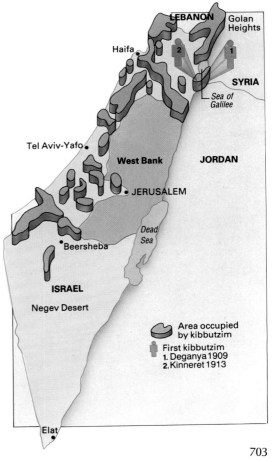

LEBANON
Golan Heights
Haifa
2
1
SYRIA
Sea of Galilee
Tel Aviv-Yafo
West Bank
JORDAN
JERUSALEM
Dead Sea
Beersheba
ISRAEL
Negev Desert
Area occupied by kibbutzim
First kibbutzim
1. Deganya 1909
2. Kinneret 1913
Elat

Italy

The country of Italy sits on a boot-shaped peninsula. It is a land of mountain ranges, rolling hills, and bustling cities—and a people of abundant spirit, energy, and optimism.

Italy is known for its rich cultural heritage and natural beauty. Its cities have spectacular churches and large central plazas. Their museums contain some of the world's best-known art. The countryside has warm, sandy beaches; high, glacier-topped mountain peaks; and rolling hills covered with green fields and vineyards.

Italy got its name from the ancient Romans. The Romans called the southern part of the peninsula *Italia*, meaning *land of oxen* or *grazing land*.

For hundreds of years, Italy has strongly influenced the history of Western civilization. One of the world's greatest empires, the Roman Empire, began with a small community of shepherds on the hillsides of central Italy. In the early 1300's, the city-states of Italy were the birthplace of the Renaissance, a great cultural movement during which many European scholars and artists studied the learning and art of ancient Greece and Rome.

Everywhere in Italy, there are reminders of the glories of its ancient past. Many monuments built by the ancient Romans still stand. In the city of Rome, art museums display the masterpieces of such Renaissance artists and sculptors as Michelangelo, Leonardo da Vinci, and Botticelli.

Yet for the Italians, time marches on, as it does for all of us. A six-lane highway now encircles the Colosseum, ancient Rome's greatest architectural wonder. Today, the splendid Roman baths provide a background for operas. In the footsteps of their ancestors, the Italians have struggled on, creating one of the most highly industrialized nations in Europe.

The country boasts several world-famous cities. Rome, the capital and largest city of Italy, was the center of the Roman Empire 2,000 years ago. Florence was the home of many artists of the Renaissance. Venice, with its intricate canal system, attracts tourists from all over the world.

More than two-thirds of Italy's people live in the cities today, mainly working in factories and offices. And industrial plants are now as much a part of the Tuscan countryside as its legendary olive groves. But even so, Italy has a timeless quality that seems to fill the heart of every visitor.

Italy Today

Italy is an ancient land but a young nation. It became a united country in the relatively recent year of 1861, when numerous smaller kingdoms on the Italian Peninsula came together under the leadership of King Victor Emmanuel II. In 1922, Benito Mussolini became premier of Italy, and by 1925, he ruled as a dictator.

After Mussolini was overthrown by King Victor Emmanuel III in 1943, Italy returned briefly to being a monarchy. Then, in 1946, the Italian people chose a republic to replace the monarchy. In 1948, a new Constitution proclaiming "a democratic republic founded on work" went into effect.

During the 1950's, Italy was transformed from an agriculture-based economy to an industry-based economy. The growth of industry was so rapid and dramatic that industrial production had more than doubled its prewar level by the 1960's.

Government bureaucracy

Italy has some political problems, mainly caused by an inefficient government bureaucracy. Previously, Italy's elections were based on *proportional representation,* so the number of votes received by a political party's candidates determined the percentage of seats in Parliament held by each party. In 1993, Italy changed its laws so that three-fourths of Parliament is elected directly and one-fourth by proportional representation.

The Christian Democrats held the office of the premier between 1987 and 1992. In 1993, many Christian Democrats were accused of corruption and colluding with the Mafia. Eleven members of Parliament, including former Premier Giulio Andreotti, were indicted. Members of the Socialist Party, including former Prime Minister Bettino Craxi, were also accused.

In 1994, voters elected a new coalition government headed by conservative business leader Silvio Berlusconi. But Berlusconi himself was investigated for allegedly bribing tax collectors while he was in business. He resigned in December 1994 and was replaced by Lamberto Dini.

Terrorism in the 1970's

During the 1970's, the Christian Democrats began to cooperate more fully with the Communists in an effort to strengthen the nation's economy and fight inflation.

This cooperation was strongly opposed by many Christian Democrats, as well as by some of Italy's allies. In addition, a leftist terrorist group called the Red Brigades created fear and disruption throughout Italy by bombing public places and killing business executives and government officials.

FACT BOX

COUNTRY

Official name: Repubblica Italiana (Italian Republic)
Capital: Rome
Terrain: Mostly rugged and mountainous; some plains, coastal lowlands
Area: 116,306 sq. mi. (301,230 km²)

Climate: Predominantly Mediterranean; Alpine in far north; hot, dry in south
Main rivers: Po, Arno, Tiber
Highest elevation: Monte Bianco (Mont Blanc), 15,771 ft. (4,807 m)
Lowest elevation: Mediterranean Sea, sea level

GOVERNMENT

Form of government: Republic
Head of state: President
Head of government: Prime minister
Administrative areas: 20 regioni (regions)
Legislature: Parlamento (Parliament) consisting of the Senato della Repubblica (Senate) with 315 members serving five-year terms and the Camera dei Deputati (Chamber of Deputies) with 630 members serving five-year terms
Court system: Corte Costituzionale (Constitutional Court)
Armed forces: 265,500 troops

PEOPLE

Estimated 2002 population: 57,092,000
Population growth: 0.09%
Population density: 491 persons per sq. mi. (190 per km²)
Population distribution: 90% urban, 10% rural
Life expectancy in years: Male: 76 Female: 82
Doctors per 1,000 people: 5.9
Percentage of age-appropriate population enrolled in the following educational levels: Primary: 102* Secondary: 95 Further: 47
Languages spoken: Italian (official) German French Slovene

The Republic of Italy includes the boot-shaped Italian Peninsula as well as the islands of Sicily and Sardinia. The independent states of San Marino and Vatican City also lie within Italy's borders. Vatican City is located within the city of Rome.

Religions:
Predominately Roman
  Catholic
Protestant
Jewish
Muslim

Enrollment ratios compare the number of students enrolled to the population which, by age, should be enrolled. A ratio higher than 100 indicates that students older or younger than the typical age range are also enrolled.

TECHNOLOGY

Radios per 1,000 people: 878

Televisions per 1,000 people: 494

Computers per 1,000 people: 179.8

ECONOMY

Currency: Euro

Gross national income (GNI) in 2000: $1,163.2 billion U.S.

Real annual growth rate (1999–2000): 2.9%

GNI per capita (2000): $20,160 U.S.

Balance of payments (2000): -$5,670 million U.S.

Goods exported: Engineering products, textiles and clothing, production machinery, motor vehicles, transport equipment, chemicals, food, beverages, tobacco, minerals and nonferrous metals

Goods imported: Engineering products, chemicals, transport equipment, energy products, minerals and nonferrous metals, textiles and clothing, food, beverages, tobacco

Trading partners: European Union, United States

In 1978, the Red Brigades kidnapped and murdered Aldo Moro, a former premier who had supported cooperation between the Christian Democrats and the Communists. Reaction to this murder led to a renewed drive against the terrorists, and hundreds of people were arrested and convicted. By the late 1980's, the wave of terrorism had ended.

In recent years, the political influence of the Roman Catholic Church has weakened in Italy. In spite of opposition from the church, divorce was legalized in 1970, and abortion was legalized in 1978. In 1985, a 1929 agreement that had made Roman Catholicism the state religion was dissolved.

Environment

The mainland of Italy is a long, mountainous peninsula extending 708 miles (1,139 kilometers) from its northernmost point to the Mediterranean Sea. The Alps form the nation's northern border, separating Italy from France, Switzerland, Austria, and Slovenia.

Western Italy faces the Ligurian and Tyrrhenian seas, while the coast of southern Italy—the "sole" of its famous "boot"—opens to the Ionian Sea, and Italy's east coast lies on the Adriatic Sea.

The landscape of Italy is varied and beautiful. Steep cliffs plunge into the sea on the southeast coast, while inland the rolling countryside features mile after mile of green fields and vineyards.

Italy is located in the Alpide belt, which cuts across southern Europe and Asia. It is one of the world's most geologically active regions. Along this belt lie the edges of the giant *plates* that make up the earth's crust. As the plates move slowly and continuously, the rocks at their edges are stretched and squeezed. This squeezing and stretching causes earthquakes and sometimes volcanoes.

Italy has the only active volcanoes in Europe. Stromboli, in the Tyrrhenian Sea, is constantly active, but violent eruptions are rare because the lava flows freely, instead of building up internal pressure. Mount Etna, on the island of Sicily, has erupted at least 260 times since its first recorded eruption in about 700 B.C.

Vesuvius, rising on the Bay of Naples in southwestern Italy, is probably the most famous volcano in the world. In A.D. 79, Vesuvius erupted and destroyed the ancient city of Pompeii.

A mountainous land

Most of Italy consists of mountains and hilly regions. The Alpine Slope region, which stretches across the northernmost part of Italy, includes huge mountains and deep valleys. The lower slopes are lined with forests. The high mountaintops consist of only barren rocks and glaciers.

The Apennine Mountains extend almost the entire length of Italy, forming a "backbone" for the country. The northern Apennines are lush and green with some of

The columns of an ancient temple rise in a meadow at Paestum. Paestum was founded about 600 B.C. by Greek settlers, who called it Poseidonia. It became a Roman colony, and received its present name, in 273 B.C.

Cypress trees line a gravel road in the hill country of Tuscany, *above.* According to legend, Cupid's arrows were made of cypress wood. Cupid was the Roman god of love, and anyone shot by his arrows supposedly fell in love.

Scaligeri Castle, *right,* once a fortress of the rulers of Verona, stands on a narrow peninsula that juts out into Lake Garda in the south-central part of the Alpine Slope.

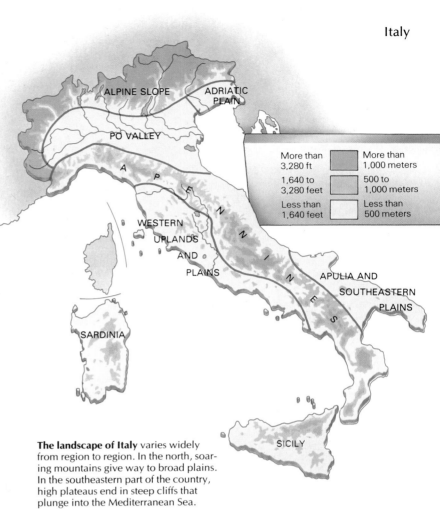

The landscape of Italy varies widely from region to region. In the north, soaring mountains give way to broad plains. In the southeastern part of the country, high plateaus end in steep cliffs that plunge into the Mediterranean Sea.

the largest forests in the country and much pastureland, while the central range supports farmland and grasslands. The southern Apennines are made up mainly of plateaus and high mountains, with few natural resources.

Fertile plains

Between the Alps and the Apennines lie the vast plains of the Po Valley, Italy's richest and most modern agricultural region. Waters from the Po River, the country's largest waterway, have created this fertile farmland.

The Western Uplands and Plains are second only to the Po Valley in agricultural importance. The northern part of the region includes the rich hill country of Tuscany and Umbria, where grain crops flourish and livestock graze. In the southern part of the region, the warm climate is ideal for growing apricots, cherries, lemons, peaches, vegetables, and wine grapes.

Apulia and the Southeastern Plains—a semi-arid limestone plateau surrounded by fertile plains—form the "heel" of Italy's "boot." This region produces most of Italy's olive oil. Water for the area's olive trees comes from the Sele River by way of the 143-mile (230-kilometer) Apulian Aqueduct, built between 1905 and 1928.

Island regions

Like the mainland of Italy, the island of Sicily has a varied landscape of mountains and plains. Mount Etna dominates the northeastern side of the island. Severe soil erosion, caused in part by the clearing of forests, has made agriculture difficult on Sicily.

Sardinia, with a landscape consisting of mountains and plateaus, lacks good farmland. Artichokes, cereals, and grapes are grown only on the island's narrow coastal plains.

Early History

About 800 B.C., a people known as the Etruscans arrived in the coastal plains of northern Italy between the Arno and Tiber rivers. The Etruscans had the most advanced civilization in Italy, and in about 600 B.C., they took control of Rome and other towns in Latium, a region south of Rome.

Under the Etruscans, Rome grew from a community of farmers and shepherds to a wealthy trading center. In less than 100 years, the Romans became so powerful that they were able to drive out the Etruscans.

The Roman Republic

Over time, a republican government began to develop in Rome. Meanwhile, Rome was extending its control to cover the rest of the Italian Peninsula. Protection and limited Roman citizenship were offered to the people who were defeated by the Romans. In return, the Romans took soldiers and supplies from the conquered cities.

Marching on to victory in city after city, the Roman army soon took to the seas to conquer distant territories. Their triumphs

marked the beginning of one of the world's mightiest empires.

The building of an empire

In the three Punic Wars fought between 264 B.C. and 146 B.C., Rome defeated the Carthaginians of North Africa and gained control of the Mediterranean Sea. Later, the Roman army conquered Greece and Macedonia.

The army's military conquests brought great riches to the Roman Empire, including tax revenues from conquered countries and the looted property seized in many wars. But the large numbers of slaves brought from conquered lands to Rome to work on the plantations created unemployment among the Romans and drove out the small farmers. As the gap widened between the rich and the poor, discontent grew among the Roman people.

Soon Rome was involved in great political turmoil. Conflicts among its leaders resulted in a series of civil wars. In 60 B.C., a three-man political alliance called the First Trium-

The Forum, a group of temples, palaces, and other buildings, *top,* served as ancient Rome's administrative, legislative, and legal center. Romans went to the Forum to hear famous orators speak and to see the treasures seized from conquered lands.

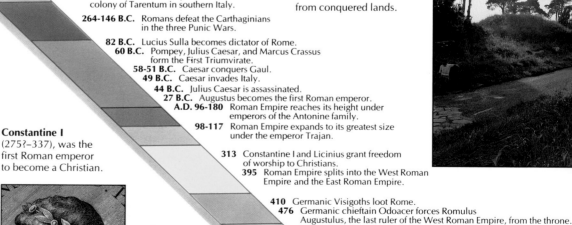

c. 800 B.C. Etruscans settle in northern Italy.
753 B.C. According to legend, Rome is founded by the twin brothers Romulus and Remus.

600's B.C. Greeks settle in southern Italy and Sicily.
c. 600 B.C. Etruscans gain control of Rome and its neighboring towns in Latium.

509 B.C. Romans drive out the last Etruscan king and establish the Roman Republic.
275 B.C. Romans gain control over most of the Italian Peninsula after a victory over the Greek colony of Tarentum in southern Italy.

264-146 B.C. Romans defeat the Carthaginians in the three Punic Wars.

82 B.C. Lucius Sulla becomes dictator of Rome.
60 B.C. Pompey, Julius Caesar, and Marcus Crassus form the First Triumvirate.
58-51 B.C. Caesar conquers Gaul.
49 B.C. Caesar invades Italy.
44 B.C. Julius Caesar is assassinated.
27 B.C. Augustus becomes the first Roman emperor.
A.D. 96-180 Roman Empire reaches its height under emperors of the Antonine family.
98-117 Roman Empire expands to its greatest size under the emperor Trajan.

313 Constantine I and Licinius grant freedom of worship to Christians.
395 Roman Empire splits into the West Roman Empire and the East Roman Empire.

410 Germanic Visigoths loot Rome.
476 Germanic chieftain Odoacer forces Romulus Augustulus, the last ruler of the West Roman Empire, from the throne.
553 Roman Empire is temporarily reunited under Emperor Justinian.
572 Lombards invade Italy.

774 Charlemagne defeats the Lombards.

800 Pope Leo III crowns Charlemagne emperor of the Romans.

962 Otto the Great, king of Germany, is crowned emperor of what later became the Holy Roman Empire.

Constantine I (275?-337), was the first Roman emperor to become a Christian.

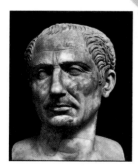

Julius Caesar, (100?-44 B.C.), *left,* was one of ancient Rome's greatest statesmen and generals.

Hadrian (A.D. 76-138), *far left,* became emperor in A.D. 117.

The Roman Empire, *above,* reached its greatest size under the reign of Trajan, a military leader who became emperor in A.D. 98 and ruled until 117. At its height, the Roman Empire included much of Europe, North Africa, and Asia Minor.

MAXIMUM EXTENT OF ROMAN EMPIRE AD117

Archaeologists reconstructed these bodies of victims of the A.D. 79 Mount Vesuvius volcanic eruption by carefully pouring plaster into the *shells* (molds) preserved in the hardened ash.

Sheep graze near the Appian Way, the first and most important highway built by the ancient Romans. Still in use today, the road first covered 132 miles (212 kilometers), linking Rome with some of its earliest conquered territories. Later, it was extended another 234 miles (377 kilometers) to Italy's southeastern coast.

virate was formed by Pompey, Julius Caesar, and Marcus Crassus.

Then, when Caesar conquered Gaul, other Roman leaders grew fearful of his power and ambition, and they ordered him to give up his command. Instead, Caesar left Gaul and invaded Italy in 49 B.C. By 45 B.C., Caesar had defeated his political enemies, and he became sole ruler of the Roman world.

A year later, Caesar was assassinated by a group of aristocrats who hoped to revive the Roman Republic. After another civil war, Caesar's adopted son and heir, Octavian, became the first Roman emperor. He took the name *Augustus,* meaning ex*alted.*

An era of peace

Under the reign of Augustus, the Roman Empire entered a period of stability and prosperity, known as the *Pax Romana* (Roman peace), that lasted about 200 years. But after Augustus died in A.D. 14, a series of *dynasties* (rulers of the same family) governed the empire. The central government in Rome could no longer hold the empire together, and provincial governors and army commanders began to declare themselves emperors.

Diocletian, a general who was named emperor by his troops in 284, divided the empire into east and west sections, each ruled by a different emperor. In 395, the empire was split into the West Roman Empire and the East Roman Empire. But the West Roman Empire could not withstand the attacks of the Germanic peoples from the north. In 476, the Germanic chieftain Odoacer forced Romulus Augustulus, the last ruler of the West Roman Empire, from the throne.

By the mid-500's, the Christian popes who now ruled Rome had acquired great influence in religious and political matters throughout the Italian Peninsula. With the help of the Frankish king, Pepin the Short, and his son, Charlemagne, the popes defeated the Lombards. In 754, they established political rule in central Italy over what became known as the Papal States.

Modern History

During the 1000's, the cities of Italy were ruled by what was later known as the Holy Roman Empire. Because the emperors lived in Germany, they had little direct power over the Italian lands. As a result, some Italian cities, such as Florence, Genoa, Milan, Pisa, and Venice, gradually developed into nearly independent *city-states*.

The Renaissance

Cultural life in the city-states helped encourage the growth of the cultural movement known as the Renaissance. Under the sponsorship of the Italian cities, painting, sculpture, and architecture reached great heights.

The word *Renaissance* comes from the Latin word *rinascere* and refers to the act of being reborn. Renaissance scholars and artists wanted to recapture the spirit of the ancient Greek and Roman cultures in their own artistic, literary, and philosophic works.

The Renaissance thus represented a rebirth of these ancient cultures.

Some Renaissance philosophers—known as *humanists*—blended a concern for the history and actions of human beings with religious concerns. The humanists studied languages, literature, history, and ethics, believing that these subjects would help them better understand the problems of humanity. They used the culture of the ancient Greeks and Romans as a model for how they should conduct their lives.

Renaissance thinking soon spread to other European countries, and Italian styles influenced nearly every area of European activity. Italy soon became attractive to foreign conquerors. King Charles VIII of France marched into Italy in 1494, and the city-states could not hold back the French army. Charles soon withdrew, but he had shown that the cities of Italy could be conquered because they were not united.

Thirteen towers built during the Middle Ages rise above the rooftops of San Gimignano, *top,* a hill town in Tuscany. More than 70 such towers were built in San Gimignano during that period, some as defense towers.

SAVOY (to France 1860)
LOMB
KINGDOM (
SARDINIA
NICE (to France 1860)
PA
18
CORSICA (to France 1768)

1000   Italian cities begin growing into independent city-states.
1000's   Normans conquer southern Italy and Sicily, and later unite them to form the Kingdom of Two Sicilies.
1215   Frederick II becomes emperor of the Holy Roman Empire.
c. 1300   The Renaissance begins in Italy.
1309–1377   Pope moves the papacy to Avignon, France.
1494   King Charles VIII of France marches into Italy.
1519   Charles I of Spain becomes emperor of the Holy Roman Empire.
1521-1559   The forces of Spain and the Holy Roman Empire defeat France in a series of wars over the control of Italy.
1700's   Austrian Habsburgs control most of Italy.
1796   Napoleon Bonaparte drives Austrians from Italy and seizes the country.
1814-1815   The Congress of Vienna returns Italy to its former rulers after Napoleon is defeated.
1848-1849   Italian states revolt against Austrian rule; Austrians crush the rebellion.
1861   The Kingdom of Italy is formed.
1866   Venetia becomes a part of the Kingdom of Italy.
1870   Rome becomes part of Italy.
1871   Rome becomes the capital of Italy.
1882   Italy becomes a part of the Triple Alliance, along with Austria-Hungary and Germany.
1911-1912   Italy occupies Libya.
1915-1918   Italy fights on the side of the Allies in World War I (1914-1918).
1922   Benito Mussolini becomes premier of Italy.
1925   Mussolini rules as dictator of Italy.
1929   Lateran Treaty establishes normal relations between the Roman Catholic Church and the Italian government.
1936   Mussolini and German dictator Adolf Hitler sign an agreement called the Rome-Berlin Axis, which outlines a common foreign policy for Germany and Italy.
1940   Italy enters World War II (1939-1945) on the side of Nazi Germany.
1943   Italy surrenders to the Allies.
1945   Mussolini is executed by anti-Fascist Italians.
1946   The Republic of Italy is established.
1947   Italy's new Constitution is adopted.
1950   Italy becomes a founding member of the North Atlantic Treaty Organization (NATO).
1958   Italy helps establish the European Economic Community.
1978   Terrorist Red Brigades kidnap and kill Aldo Moro, a former premier who was expected to become Italy's next president.

Lorenzo the Magnificent (1449-1492), *left,* headed the Medici, a ruling family of Florence during the Renaissance.

Giuseppe Garibaldi (1807-1882) fought to unite Italy into a single kingdom.

Leonardo da Vinci (1452-1519), *far left,* was one of the greatest painters of the Renaissance.

The huge dome of the Cathedral of Santa Maria del Fiore dominates the skyline of Florence. The dome's interior is decorated by *Last Judgment*, a *fresco* (wall painting) by Giorgio Vasari and Federico Zuccari. The cathedral, begun in 1296, is one of the finest examples of architecture of its period in Tuscany.

Modern Italy began to take shape in 1859, as parts of the Italian Peninsula joined the Kingdom of Sardinia. The territories of the Kingdom of Sardinia included northwest Italy and the island of Sardinia. The dates on the map show when each region joined the kingdom.

Invaders and conquerors

In 1519, Charles I of Spain, a member of the Habsburg family, became emperor of the Holy Roman Empire. His troops looted Rome in 1527, and by 1559 they had seized Milan and Sicily from France. Then, as Spanish influence weakened, control of Italy passed from the Spanish Habsburgs to the Austrian Habsburgs.

In 1796, the French ruler Napoleon Bonaparte led his army into Italy. French control lasted less than 20 years, but in that time, Napoleon introduced democratic reforms to the Italian people. Italy's former rulers were reestablished after Napoleon's defeats. But the Italian people, impressed by the ideals of the French Revolution, had begun to dream of a united, independent Italy.

Birth of a nation

In 1848, the Italian kingdoms revolted against their rulers, but Austria crushed the revolutions in 1849. After many years of fighting, the Italians drove out the Austrians. In 1861, Victor Emmanuel II, supported by a nationwide vote, declared the formation of the Kingdom of Italy and became Italy's first king.

In 1866, the northwestern region of Venetia also became part of Italy. Only Rome, which remained under the pope's control, and tiny San Marino were not part of the new kingdom. By 1870, the pope's territory was reduced to the boundaries of the Vatican, and Rome became the capital of Italy in 1871.

After World War I (1914-1918) ended, many workers were unemployed and the Italian people became greatly discontented with their government. Soon, a movement called *Fascism*, led by Benito Mussolini, gained power. The Fascists favored strict government control of labor and industry. By 1925, Mussolini ruled as dictator.

In 1940, Mussolini led Italy into World War II (1939-1945) on the side of Nazi Germany. After suffering a series of crushing defeats, Italy surrendered in 1943, and Mussolini was executed by anti-Fascist Italians in 1945. Italy was briefly ruled by a king before it became a republic in 1946.

Rome

Almost 3,000 years ago, a group of shepherds built a small village on a hill along the banks of the Tiber River. From these simple beginnings came the great city of Rome, the center of Western civilization for more than 2,000 years.

Today, Rome is one of the most beautiful and historic cities in the world. From its ancient monuments to its priceless works of art, its *piazzas* (squares) and open-air markets, the city is a treasure-trove of Western culture. Because of its long history, Rome is known as the *Eternal City*.

The city stands on about 20 hills, which include the famous seven hills on which ancient Rome was built—the Aventine, Caelian, Capitoline, Esquiline, Palatine, Quirinal, and Viminal hills. The Tiber River flows through the center of Rome to the Tyrrhenian Sea.

Ancient beginnings

According to legend, Rome was founded by twin brothers, Romulus and Remus, who were heirs to the throne of the ancient Italian city of Alba Longa. Soon after their birth, their uncle, who wanted to remain king, had Romulus and Remus put into a basket and thrown into the Tiber River. Legend says that a female wolf found the twins and nursed them. Then they were rescued by a shepherd, who, with his wife, raised the two boys.

When they were young men, Romulus and Remus set out to found a city of their own. To settle a quarrel about the site of the proposed city, they decided that whoever counted the most vultures in flight would select the site. Romulus saw 12 vultures, while Remus saw only 6.

But Remus, suspecting that Romulus had cheated, mocked his brother. For this act of disloyalty, Remus was killed. Romulus then named the new city after himself and became the city's first ruler.

The story of Romulus and Remus and its bloody ending are part of Roman mythology, but the violence upon which much of Rome was built was quite real. The ancient Romans conquered lands through military conquests, seized the riches they found there, and took the conquered citizens as slaves.

And the killing was not confined to the battlefield. From A.D. 80 until the 400's, the Colosseum was the scene of bloody battles staged for the entertainment of the Roman people. The slaughter included combat between gladiators, who were trained warriors, as well as between men and wild animals. On the day the Colosseum opened, 5,000 wild animals were slaughtered in its arena.

But in spite of its cruelty, the Roman Empire gave the Western world the foundations of its modern governmental, legal, and military systems, as well as its language and arts. Present-day Rome abounds with reminders of its ancient citizens—not only of their flaws but also of their triumphs.

Historic sites

Not far from the Roman Forum, the seat of ancient Rome's government, stands the Forum of Trajan. At its center, a column 100 feet (30 meters) high bears a striking series of sculptures depicting scenes from Trajan's wars. Sadly, Trajan's column—like many of Rome's ancient monuments—must now be protected under a cover of sheeting from damage caused by air pollution.

The ruins of the Colosseum, *above,* still rank among the finest examples of Roman architecture and engineering. This theater, in use from A.D. 80 until the 400's, seated about 50,000 spectators on marble and wooden benches.

The Piazza Navona follows the oval form of the ancient Circus of Diocletian. Across from the Fountain of Neptune, *right,* stands the Church of Sant' Agnese in Agone, partly rebuilt in the 1600's. During the 1700's, the center of the piazza was flooded as part of a local festival.

Romans and tourists alike relax on the Spanish Steps, one of the city's most popular meeting places. It is one of the many splendid sights of Rome, capital of Italy and once the center of Western civilization. Unlike most leading cities, Rome has little industry, and its economy depends chiefly on tourism.

Underneath the streets of Rome lie the catacombs—underground passages cut into the rocks by the early Christians in the 200's and 300's. During the period of Roman persecution, the Christians took refuge in the catacombs.

Visitors to Rome have two favorite meeting places—the Fountain of Trevi and the Spanish Steps. According to legend, a visitor who tosses a coin into the fountain is assured of a happy return to Rome someday.

The 200-year-old Spanish Steps slope gently upward to a square dominated by the Church of Trinità dei Monti. From the Spanish Steps, tourists can join the local people in the typical Roman pastime of *dolce far niente* (delightful idleness) and simply watch the world go by.

A wealth of ancient monuments and historic ruins has kept alive the magic of Rome's long and colorful past. Famous landmarks include the Colosseum (14), St. Peter's Church (1), the Spanish Steps (9), the Palace of Justice (5), the Circus Maximus (17), and the Roman Forum (12). Another sight of historic interest is the Baths of Caracalla, which served both as public baths and a luxurious meeting place in ancient Rome.

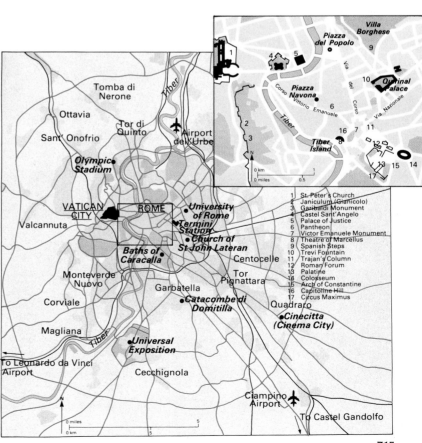

1 St. Peter's Church
2 Janiculum (Granicolo)
3 Garibaldi Monument
4 Castel Sant'Angelo
5 Palace of Justice
6 Pantheon
7 Victor Emanuele Monument
8 Theatre of Marcellus
9 Spanish Steps
10 Trevi Fountain
11 Trajan's Column
12 Roman Forum
13 Palatine
14 Colosseum
15 Arch of Constantine
16 Capitoline Hill
17 Circus Maximus

People

Almost all of the people who live in Italy today belong to the same Italian ethnic group. Often noted for their black hair, olive skin, and dark eyes, the Italian people are descendants of the Ligurian peoples, the original inhabitants of the peninsula.

As different groups settled the land, they created some regional variations that can still be seen throughout Italy. The Arabs and Normans who once occupied southern Italy and Sicily left their mark. People of Greek and Albanian heritage can be found in Sicily, as well as in Calabria. In Friuli-Venezia Giulia, near the Slovenian border, many people have Slavic characteristics.

The largest ethnic minority is the Germans, who live in the Trentino-Alto Adige region bordering Austria. German is the first language of many people in this region.

Slovenes, another ethnic group, live in the Trieste area, along the border of Italy and Yugoslavia, and speak *Slovene*, a Slavic language. A number of ethnic French people live in the Valle D'Aosta region, near Italy's border with France and Switzerland. Slovene and *Ladin*, a language similar to the Romansch of the Swiss, are spoken in northern Venetia.

Overall, the people of Italy have much in common, including their language, their religion, and a national character built on the importance of the family. The Italian family's vast social network of relatives and friends reflects the strength of a well-knit society.

Their religious faith and their family ties are a source of strength for the Italian people. In many ways, these basic values have helped the Italian people through the transformations their country experienced during the 1900's.

Language and dialects

Italian is the official language of Italy. An increasing number of Italians now choose standard Italian, which developed from a dialect spoken in Tuscany, over their regional dialect. The standardization of language in Italy today is due partly to television, radio, books, and education, and even to travel and military service.

Nevertheless, many dialects are still spoken throughout the country. In the northern region, these include Emilian, Ligurian, Lombard, Piedmontese, and Venetian. In central Italy, some people speak Corsican, Roman, Tuscan, or Umbrian. And in the south, the dialects are Abruzzese, Apulian, Calabrian, Neapolitan, and Sicilian.

Religion

About 95 per cent of the Italian people are Roman Catholic, and the Roman Catholic Church has had a tremendous influence on the country's political and social life. Although church influence has weakened in

A sidewalk cafe is a popular meeting place for a group of friends in a Tuscan town, *below*. A friendly chat over a glass of wine or a dish of *gelato* (creamy Italian ice cream) is a typical Italian way of strengthening relationships.

Members of a religious community enjoy a *passeggiata* (stroll) through the Italian countryside. Families generally enjoy a stroll in the early evening as a way of relaxing, getting some exercise, and visiting with friends.

A woman in Rome hangs laundry outside her apartment window to dry. Most city-dwellers live in concrete apartment buildings, and many buy, rather than rent, their apartments. A few wealthy people live in single-family homes.

recent years, religion still plays an important role in the lives of many Italians. While only about 30 per cent of the Italian people attend church regularly, most are baptized, married, and buried with Roman Catholic rites.

Local festivals celebrating religious holidays are held in Italy throughout the year. During the first week of January, many communities sponsor pageants re-creating the arrival of the Magi, the three wise men who followed the star of Bethlehem when Jesus Christ was born. On Good Friday in Taranto, a procession of hooded *penitents* (people sorry for their sins) makes a 14-hour journey around the city.

Ascension Day in Cocullo, in the Abruzzi region, is celebrated with a procession in which pilgrims and singers carry live snakes in their hands. Perhaps the most spectacular festival is the annual Carnival, held during the 10 days preceding Lent (the 40 weekdays between Ash Wednesday and Easter).

Family life

The maintaining of close family ties has always been an important tradition in Italy. The family has its strongest roots in the small farming communities that existed throughout the nation before the coming of industrialization.

Most Italians no longer live in the farms and villages of their ancestors. Hoping to find a better life, many people moved to the cities. But the family values that were passed down through the generations remain strong.

A band of accordion players serenades a woman in Calabria, a region in southwestern Italy. The Italians have a great love of music and a rich musical heritage. The first operas were composed in Florence in the 1590's.

Foods of Italy

To many people, Italian food means—simply—pizza! The word *pizza* means *pie* in Italian, and the dish was invented in Naples in the 1700's. *Pizza napoletana verace* (true Neopolitan pizza), which consists of tomatoes, garlic, olive oil, and oregano baked on dough over a wood fire, can still be purchased in the tiny street stalls of Naples—though electric ovens have replaced the wood fires.

But for all its popularity, pizza is just one of the many delicious creations served throughout Italy. The abundance of fresh fruits, vegetables, herbs, and spices grown in the rich soil of Italy provide the basis for *la cucina Italiana* (Italian cooking).

The Italian people seem to enjoy eating as much as they enjoy cooking. An Italian meal is a festive, social occasion as well as a way to fill an empty stomach.

An Italian meal

Families gather together to eat their main meal at midday. The first items on the family table are *antipasti* (appetizers)—cold meats and vegetables that might include salami, olives, and artichoke hearts. Next come *primi piatti* (first courses)—pasta, soup, rice, or *polenta* (cornmeal porridge).

Pasta—a mixture of wheat flour, oil, and water—appears in one form or another on every Italian table. It is an ancient food, even enjoyed by the Etruscans thousands of years ago.

Pasta comes in a great variety of shapes. Long, slender varieties of pasta include *capelli d'angelo* (angel hair) and *vermicelli* (little worms). *Ziti* and *rigatoni* are short, round, hollow pastas. Long, flat pastas include *linguine* (little tongues) and *lingue di passero* (sparrow tongues).

Wheat-flour pasta and olive oil are the foundations of southern Italian food. The north is better suited to grazing livestock, so butter is used more often than olive oil. In the north, rice or polenta is a more common dish than pasta. *Risotto alla milanese* (saffron rice), for example, comes from the northern city of Milan.

After the first courses come *secondi piatti* (second or main courses) of meat, poultry, fish, and eggs. Italy is noted for its cured

A woman sells fresh spinach at an outdoor market in Rome, *above*. Many spinach dishes are described as *alla florentine* (in the Florentine style). Spinach is also used to enrich pasta dough, and it gives the pasta a green color.

A gondolier looks on as diners enjoy a meal at one of Venice's many sidewalk cafes. Italians take great pride in the quality of their food, whether cooked at home or in a *trattoria*—a small, informal family restaurant.

meats and sausages, such as *prosciutto* (Parma ham) and *mortadella* (a Bolognese sausage of heavily spiced pork). Fresh meats may be served *alla griglia* (grilled or charcoal-broiled) or *arrosto* (roasted).

Italians enjoy fish of all kinds, including such exotic varieties as *pesce spada* (swordfish) and *calamari* (squid). Such famous specialties as *fritto misto di mare* (fried fish and shellfish) and *zuppa di pesce* (fish soup) include several varieties of fish.

Contorni (vegetables and salads) are next on the Italian menu. When tomatoes were brought to Europe from the New World in the 1500's, Italian cooks greeted them with enthusiasm. Other vegetables include *carciofi* (artichokes) and *piselli* (baby peas).

Dolci (sweets) are the final course of an Italian meal. The best-known Italian dessert, *gelato* (creamy ice cream), may be served with whipped cream, or *affogato* ("drowned" in whisky). Also popular is *granita*—ice crystals served with fruit syrup.

A Sardinian dessert, *sebadas,* consists of cheese-filled, fried ravioli covered with wild honey. Turin is famous for its fine chocolates, especially the hazelnut-flavored *gianduiotti.* Another traditional dessert is *Baci di Dama* (Lady's Kisses)—a blend of almonds, flour, butter, sugar, liqueur, vanilla, and chocolate.

The finishing touch

After their meal, Italians enjoy coffee, and their imaginative ways with that beverage are appreciated throughout the world. *Espresso,* a strongly flavored coffee drink, is popular, as well as *cappucino,* in which steamed, foaming milk is added to espresso. Milder coffee drinks include *caffe lungo* (with added water) and *caffe macchiato* (espresso with milk).

A pasta-maker shapes the dough—a "paste" of flour, oil, and water—for Italy's staple food. Pasta ranges from long, thin *spaghetti* to short, fat *macaroni.* Homemade pasta is usually cooked and eaten as soon as it is made.

A fish market at Venice's Ponte di Rialto displays the day's catch. Fish from the Mediterranean Sea are an important part of Italian cooking. Recently, however, water pollution has lessened the quality of Mediterranean fish.

The Arts

Italy has made important contributions to the art, architecture, literature, and theater of the world since the early Middle Ages. But nowhere is the genius of the Italian masters so evident as in the paintings and sculpture of the 1400's and 1500's.

During that time, Italy was made up of about 250 individual cities. Some, like Florence, Venice, and Milan, were rich and powerful *city-states* that operated almost independently from the Holy Roman Empire. Governed by powerful families, the city-states became leading centers of art and learning.

Renaissance painters

In the 1300's and 1400's, the thriving cultural life of Italian cities such as Florence, Milan, and Venice developed into the movement known as the Renaissance. At that time, the Medici family ruled Florence and controlled one of the largest banks in Europe. With their financial backing, the city of Florence supported some of the world's greatest artists.

Michelangelo and Raphael were among the great artists helped by the Medici. When

The bright colors and bold brushstrokes of Titian's *Presentation of the Virgin, above,* are typical of Venetian art during the Italian Renaissance. Titian's style strongly influenced European painting for more than 200 years.

Bernini's chapel altarpiece, *left,* created for the Cornaro Chapel in Rome, shows an angel poised to drive an arrow into the heart of the swooning Saint Teresa. Bernini's work typifies the highly ornamental and intensely dramatic baroque style.

The palace of Duke Frederico da Montefeltro, *below,* who ruled Urbino from 1444 to 1482, is a glorious monument to the Italian Renaissance. Its library is filled with illuminated manuscripts, and its galleries display the works of Uccello, Signorelli, and Titian. In addition to the duke's living quarters, the Ducal Palace also has a stateroom, theater, two chapels, and a secret courtyard garden.

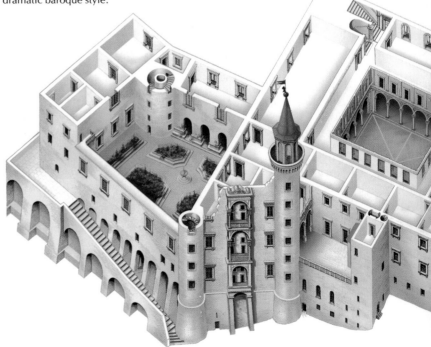

Leonardo da Vinci painted *The Annunciation,* *above,* in the 1470's for the Convento di S. Bartolomeo at Monteoliveto, just southwest of Florence. The panel now hangs in the Uffizi Gallery in Florence.

Botticelli's *Primavera,* *left,* shows his interest in beautiful mythological subjects, as well as the clear, rhythmic lines, delicate colors, and poetic feeling that distinguished his work. Botticelli lived and worked in Florence during the late 1400's.

Michelangelo was a young man, his work came to the attention of Lorenzo de' Medici, also known as Lorenzo the Magnificent. Lorenzo invited Michelangelo to stay at his palace, where the great artist began to develop the distinctive style that was to mark his work.

Raphael settled in Florence in 1504, where he studied the paintings of Leonardo da Vinci. In 1508, Pope Julius II asked Raphael to work in Rome, and there he created perhaps his greatest work—the series of frescoes that decorate the pope's private quarters in the Vatican.

Leonardo da Vinci, one of the greatest painters of the Italian Renaissance, was born near Florence, but spent much of his early career as a court artist for Lodovico Sforza, the duke of Milan. When the French overthrew Sforza in 1499, Leonardo left Milan and returned to Florence. There, he and Michelangelo were hired by the Florentine government to decorate the walls of a new hall for the city council with scenes of the city's military victories.

Italian architecture

The many beautiful churches throughout Italy stand as proof of the remarkable abilities of Italian architects. One example of the great flowering of Italian architecture that occurred during the Renaissance is the dome for the Cathedral of Florence—a masterpiece of architectural design created by Filippo Brunelleschi.

Perhaps the greatest achievement of Italian Renaissance architecture was St. Peter's Church, originally designed by Donato Bramante in the early 1500's. The building's most outstanding architectural feature is its magnificent dome, designed by Michelangelo.

Literature and music

Italian Renaissance writers produced a number of important works, including *The Prince*—a book that describes the methods a strong ruler might use to gain and keep power—written by Niccolò Machiavelli in 1513.

The Italian love of music may have found its greatest expression in the first operas, which were composed in Florence in the 1590's. The development of opera resulted from the interest of Florentine noblemen, musicians, and poets in the culture of ancient Greece. They believed that the Greeks sang, rather than spoke, their parts in Greek drama. Opera emerged as an art form during the baroque period of the 1600's and 1700's.

The Italian talent for powerful creative expression lives on. Present-day artists, like all the generations that followed the Renaissance and baroque masters, still use their gifts to create works of extraordinary visual and emotional impact. Inspired by the genius that went before them and surrounded by the masterpieces of the past, modern Italian artists carry on the brilliant tradition that began centuries ago.

Agriculture

Italy has a very long tradition of agriculture. The Etruscans, the ancient people of the Italian Peninsula, introduced sophisticated irrigation techniques into the Po Valley. Later, the Romans became skilled in the growing of fine fruits, especially wine grapes.

In present-day Italy, agriculture no longer plays a major role in the nation's economy. Agriculture, forestry, and fishing now account for only 4 per cent of the *gross domestic product* (the total value of goods and services produced within a country in a year), and employ only about 10 per cent of Italy's workers.

Although about 40 per cent of Italy's land is used for crops, farming remains a somewhat backward industry. Most of Italy's farms are individually owned and about 75 per cent of them cover less than 12 acres (5 hectares). Although some of Italy's agriculture has been modernized, much of it remains poor, especially in the south and in the mountain areas.

"Green Plans" for agriculture

After World War II, many Italians left their farms for higher-paying work in the cities. To slow the migration of workers from the land to the factories, the Italian government developed a series of nationwide "Green Plans" in the 1960's and 1970's. The Green Plans attempted to reform the land-tenure system, which had contributed to the development of small, unprofitable farms. The government plans also provided farmers with financial aid for improving agricultural techniques.

The Green Plans helped to increase the land's general productivity. They also encouraged specialization in high-quality produce, such as exotic fruits and vegetables, for both domestic use and export.

Despite the efforts of the government, agriculture has shown little growth in recent decades, and many farms are still small and unprofitable. Although the production of sugar beets, tomatoes, and other crops has increased, the nation's wheat crop declined during the 1980's. Today, although Italy is a major cereal-producing country, it imports large supplies of wheat, and much of Italy's pasta is made with imported flour.

A herd of sheep graze in the rolling pasturelands of Tuscany. In addition to grazing land, the Tuscan hillsides support much fertile farmland, including large fields of grain, olives, and grapes. The famous Chianti wine is produced in Tuscany.

A vineyard worker enjoys a lunch of pasta. The farming population of Italy is aging quickly, and many fear that the younger generation will not take the place of older farmers as they retire.

Rice is grown in the well-irrigated Po Valley, *below right,* Italy's most important agricultural region. Other cereal crops include wheat and corn. The government controls the supply of domestic wheat as well as the import of foreign wheat.

Grapes and wine

Grapes, Italy's most valuable crop, are grown throughout the country, but the fertile northern valleys produce the best crops. About 2.7 million acres (1.1 million hectares) of Italy's fertile land is used for growing grapes.

Italy is the largest producer of wine in the world, and most of its grapes are used for producing wine. Almost every region in Italy produces its own wine, for an estimated total of 4,000 to 5,000 different Italian wines.

Chianti, perhaps the most familiar Italian wine, comes mainly from Sangiovese grapes native to the regions of Tuscany and Umbria

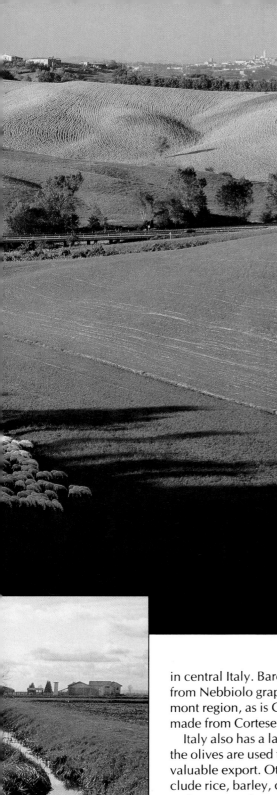

A jet of water irrigates a field in Calabria. The southern region of Italy, with its long, hot summers, occasionally suffers drought. The land is used for sheep grazing more than growing crops.

Haymaking provides a store of winter fodder and bedding for livestock in the north of Italy. This area, which includes the lower regions of the Alpine Slope, is protected from extreme cold by the Alps.

in central Italy. Barolo and Barbaresco, made from Nebbiolo grapes, are wines of the Piedmont region, as is Gavi, a crisp white wine made from Cortese grapes.

Italy also has a large olive crop. Most of the olives are used to make olive oil, another valuable export. Other important crops include rice, barley, corn, oats, and rye. Italy ranks among the world's largest producers of sugar beets, and grows more than half the world's artichokes. It also grows oranges, peaches, apples, tomatoes, and potatoes.

Livestock is raised throughout the country. Northern Italy is noted for its dairy products,

beef cattle, and its pig and poultry farms. The northern area also provides grazing for large herds of sheep and goats. The Po Valley is famous for the production of fine silk, and mulberry trees are grown there to feed the silkworms.

Coastal and deep-sea fishing in the Mediterranean employs a large number of people and enjoys a bountiful catch. Even so, the industry does not meet Italy's needs because fish is such an important part of the Italian diet.

Economy

After World War II ended in 1945, Italy shifted from an agriculture-based economy to an industry-based economy, and entered a period of tremendous economic growth. Today, northern Italy is one of the most advanced industrial areas in Western Europe.

Between 1953 and 1968, the nation's industrial production almost tripled. In 1958, Italy became one of the founding members of the European Economic Community and strengthened its economy through increased trade. The economic boom suffered a setback during the worldwide recession of the 1970's. But by the late 1980's, Italy's economy had recovered enough to place it among the world's leading industrial nations.

Industrial progress in Italy has been achieved despite the country's lack of natural resources. Italy depends heavily on other countries for its energy supply, importing more than half of its petroleum. Large amounts of natural gas from the Po Valley are piped into the cities of the north, and hydroelectric plants in the Alps provide power for northern factories. Hydroelectric plants contribute about 25 per cent of the nation's electrical supply.

The government's National Hydrocarbon Agency controls the production and distribution of Italy's petroleum and gas. The government also owns or controls much of Italy's business and industry, including steel mills, public utilities, shipbuilding companies, most of the nation's railroad network, the Alitalia airline, and the Bank of Italy.

Manufacturing

Today, about 20 per cent of the Italian work force is employed in industry. Clothing, including shoes, is the leading manufactured product in Italy. Many smaller businesses are involved in the production of textiles and clothing, shoes and leather goods, furniture, and other craft products. Italian silks and wools are prized all over the world.

Other important manufactured goods include automobiles, chemicals, electrical and nonelectrical machinery, petroleum products, and processed foods. Most heavy industry is concentrated in northwest Italy, in the triangle formed by Milan, Turin, and Genoa. Major activities in this area include the production of automobiles, household

A woman selling hand-crafted goods, *far right,* waits for customers outside her shop in Venice. Tourism greatly benefits Italy's small businesses and provides jobs for many service workers.

A fashion model shows off an Armani design at a Milan fashion show, *below.* Italian designers are international trendsetters in many fields, including fashion.

A glassblower on the island of Murano, *above,* near Venice, practices an art that has flourished in Italy since the 1200's.

Marbleworkers shape and polish stone from the famous marble quarries of Carrara, in northern Italy. Renaissance sculptors prized Carrara marble for its white color and compact grain. The stone is cut from the Alpine slopes.

The fully automated production line, *right,* at the Fiat factory in Turin turns out one of Italy's most important exports. Fiat is a leading manufacturer of mass-produced automobiles, while Alfa-Romeo makes luxury cars.

appliances, iron and steel, and machinery and machine tools.

Tourism

Although Italy is limited in the natural resources that provide raw materials for industry, it is richer than many other countries in natural beauty and historic interest. The artistic and architectural treasures of Rome, Venice, and Florence, the ruins of ancient Pompeii, and the world-class skiing on its Alpine slopes have all helped make tourism a major industry in Italy.

About 50 million tourists visit Italy each year, contributing billions of dollars to the economy. Tourism is a major part of the nation's service industries, keeping people employed in the hotels and restaurants that accommodate visitors.

Looking to the future

Despite Italy's economic achievements, some problems remain. Most postwar industrialization took place in northern Italy, leaving the south lagging far behind. Today, southern Italy has a higher unemployment rate than the north, and the percentage of people who still work in agriculture is much higher.

Italy's lack of abundant natural resources, along with the high cost of importing raw materials, limits the nation's industrial growth. And because Italy is so dependent on foreign oil, the rise in petroleum prices during the 1970's caused inflation.

Most of Italy's industrial activity is centered in the triangle formed by Milan, Turin, and Genoa in northwest Italy. Firms located there include the Fiat automobile plant, shipbuilding facilities, food processing plants, and petroleum companies.

◯ Chemicals and oil refining

⬡ Motor vehicles

◉ Iron and steel

◉ Machinery

◗ Electrical machinery

◉ Textiles and clothing

725

Venice

One of the world's most famous and unusual cities, Venice lies on a group of islands in the Adriatic Sea, 2-1/2 miles (4 kilometers) off the Italian coast. The main islands are linked to each other by more than 400 bridges across canals. The city's houses and other buildings rest on great posts driven into the mud, rather than on solid ground.

The streets of Venice have an eerie silence about them, due to the absence of automobile traffic. The dominant sounds are church bells and the footsteps of pedestrians.

"La Serenissima"

From earliest times, the Venetian economy was based on fishing and trading. By the A.D. 800's, Venice had developed into a nearly independent city-state. It traded goods with Constantinople (now Istanbul), as well as with cities on the Italian mainland and the northern coast of Africa.

In 1380, Venice defeated Genoa—its rival sea power—and gained control over trade in the eastern Mediterranean Sea. Soon, Venice became one of the world's largest cities, reaching the height of its power during the 1400's.

The city, which called itself "La Serenissima" (the Most Serene Republic), then included Crete, Cyprus, the Dalmatian coast (now part of Croatia), and part of northeastern Italy. Venetian ships carried almost all the silks, spices, and other luxury items that reached Europe from Asia.

Like the leading citizens of other Italian city-states during the Renaissance, the merchants and aristocrats of Venice used some of their great wealth to support the arts. As a result of their patronage, Venice's paintings, textiles, and handicrafts came to be known and prized throughout the civilized world.

In time, the paintings of such Venetian masters as Gentile da Fabriano and Giovanni Bellini, Vittore Carpaccio, and Giorgione took their place among the greatest works in the Western world. Meanwhile, the architects of Venice constructed magnificent palaces and churches along the city's canals.

Venice's golden age came to a halt after the Portuguese explorer Vasco da Gama dis-

High tides regularly flood Venice's walkways and alleys. Rising waters and sinking land have put the city in great danger, but government funds are now being used for projects to prevent further damage.

The Rialto Bridge crosses the Grand Canal in the heart of Venice, *below right*. The present marble causeway, built in the late 1500's, replaced a wooden bridge. Its humped shape allowed the armed Venetian ships to pass underneath.

covered a sea route to India and Christopher Columbus discovered America in the late 1400's. The center of trade in Europe then shifted to the Atlantic Ocean and the New World, and Venice—once so rich and powerful—declined in importance.

Present-day Venice

Many of the art treasures created during Venice's centuries of wealth and power can still be enjoyed by visitors today. The Basilica of Saint Mark, one of the world's outstanding examples of Byzantine architecture, stands in the southern section of the city. Its interior is richly decorated with mosaics, carvings, and colored marble.

Next to the Basilica of Saint Mark is the Doges' Palace, a huge pink-and-white Gothic building that was once the home of the rulers of Venice. The palace is joined to the state prison by a narrow, covered bridge called the *Ponte dei Sospiri* (Bridge of Sighs) for the unhappy prisoners who crossed it on their way to trial.

A view from the Basilica of Saint Mark, *left,* takes in St. Mark's Square and the island church of San Giorgio Maggiore. St. Mark's Square has many sidewalk cafes where visitors can delight in the beauty and charm of Venice over a cup of *espresso.*

A gondolier guides his passengers through the canals of Venice, *above.* Gondolas, once the city's chief transportation, are being replaced by motorboats.

A drowning city

The water that gives Venice so much of its charm also threatens to be its downfall. Floods caused by high tides during winter storms, as well as polluted air and water are now endangering the city.

In 1966, damage due to severe flooding cost millions of dollars and destroyed many of Venice's paintings and statues. The city was sinking by about 1/5 inch (5 millimeters) a year until the 1970's, when laws were passed that restricted the drainage of underground water by industry. In the late 1980's, the government approved billions of dollars for projects to protect the city from further flooding and erosion.

The city of Venice is built on about 120 islands 2-1/2 miles (4 kilometers) off the northeast Italian coast. About 150 canals connect the islands. Part of Venice extends onto the Italian mainland.

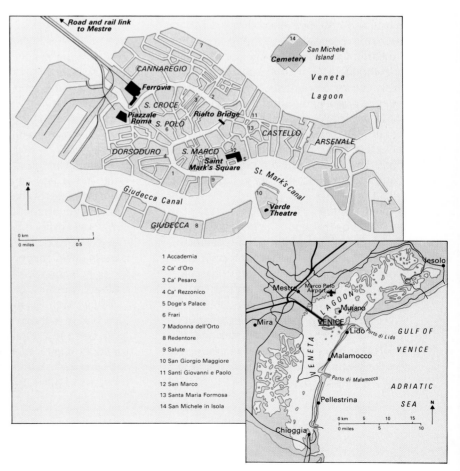

1 Accademia
2 Ca' d'Oro
3 Ca' Pesaro
4 Ca' Rezzonico
5 Doge's Palace
6 Frari
7 Madonna dell'Orto
8 Redentore
9 Salute
10 San Giorgio Maggiore
11 Santi Giovanni e Paolo
12 San Marco
13 Santa Maria Formosa
14 San Michele in Isola

Sicily and Sardinia

In addition to the mainland peninsula, the nation of Italy also includes Sicily, the largest island in the Mediterranean Sea, and Sardinia, the second largest island.

Sicily lies off the southwest coast of Italy across the Strait of Messina. Palermo, the capital of Sicily, is the island's center of industry and trade, as well as its chief seaport. A ferry links Palermo with Tunis on the coast of North Africa.

Sardinia lies off the west coast of the Italian mainland, 9 miles (14 kilometers) south of the French island of Corsica in the Tyrhennian Sea. The island of Sardinia and several small islands nearby form the region of Sardinia.

Sicily

More than 85 per cent of Sicily is covered by hills and mountains that rise to their highest point of 11,122 feet (3,390 meters) on the snow-capped peak of Mount Etna. Located on the east coast of the island, Mount Etna is one of the world's most famous active volcanoes.

Despite its violent nature, Mount Etna is very beautiful, with its high, snow-capped peaks and lush, tree-lined slopes. Colorful orchards, vineyards, and orange groves nestle around its base to complete the picture.

Although Mount Etna erupts periodically, the area surrounding the volcano is the most heavily populated region of Sicily. The volcanic ash makes the soil rich and fertile, creating excellent conditions for growing olives, grapes, citrus fruits, and cereals. At night, one can see the red glow of the volcano's lava reflected in the clouds above.

Sicilian culture

At its narrowest point, the distance between Sicily and mainland Italy is only about 2 miles (3 kilometers)—a short trip on a ferryboat. Yet the traveler who ferries to Sicily from Italy finds a completely different society.

To begin with, Sicilian dialects show traces of Arabic, as well as Greek and other European languages. These are among the marks left by centuries of foreign invasion

The harbor at Alghero in northern Sardinia is crowded with tourists taking in the sights. This charming seaside town shows the Spanish influence in the architecture of its cathedral. Many people who live in Alghero speak a Catalan dialect.

The coastline of Sardinia, *right,* has a rugged beauty all its own. But, except for its mineral deposits, the island has few natural resources. Although the island's economy relies on the tourist trade, many Sardinians resent the wealth of their visitors.

and rule—by Greek colonists as early as the 700's B.C., and later by the Carthaginians, Romans, Germanic tribes, Byzantines, Muslims, Normans, French, Germans, Spanish, and Austrians. Sicily's ancient monuments range from the Greek temples of Zeus and Hera at Agrigento to the Norman Cathedral in Palermo.

Years of foreign rule have left their influence on the Sicilian character as well as on the landscape. Some Sicilians still distrust all forms of government and tend to value personal and family honor over loyalty to their country. This code of honor is known as *omertà* (law of silence), and it forbids telling the police about crimes considered to be private affairs.

Today, Sicily is an island of poor farmers who barely manage to scrape a living off the land. Despite the Italian government's efforts to modernize farming, most Sicilian farmers still use old-fashioned methods and equipment. In addition, many people leave Sicily

for higher-paying jobs in northern Italy and other countries.

Sardinia

About 90 per cent of Sardinia is mountainous. Very few people live in the mountains because the steep slopes and heavy rainfall produce landslides and floods.

The only important lowland region is the Campidano plain of the southwest, where almonds, grapes, herbs, lemons, olives, oranges, and wheat are grown. Elsewhere, large herds of sheep and goats are grazed. Although many forests have been cleared, the northeast part of the island still has many cork oak trees, and cork is a leading export. Mines produce copper, iron, lead, lignite, manganese, silver, and zinc.

Industrial development on the island remains limited, but tourism has become an important part of the economy. In recent years, the beautiful Costa Smeralda (Emerald Coast) has become very popular due to the area's beautiful beaches and plentiful tourist accommodations.

Like Sicily, Sardinia has a long history of foreign invasion. Its capital and largest city, Cagliari, was founded by the Phoenicians. Later, the island was settled by the Romans. Sardinia was invaded by Germanic tribes in the 400's, occupied by the Byzantines in the 500's, and attacked by Arabs from the 700's to the 1000's. In 1297, Sardinia was given to the Spanish Crown of Aragon, but Aragon did not complete its conquest of the island until the end of the 1400's. In 1713, under the Treaty of Utrecht, Austria gained control of the island.

In 1720, along with the Piedmont region on the northwest coast of Italy, the island became a part of the Kingdom of Sardinia. The center of the kingdom's government was in Turin, where the first movement for a united Italy began to take shape in the 1820's and 1830's.

The view from the top of the ancient Greek theater in Taormina is perhaps Sicily's most beautiful view. The Calabrian coast can be seen in the background. The theater was rebuilt in the Roman style in the A.D. 100's.

A Sardinian worker takes a break. Most islanders are poor farmers who barely manage to scratch a living from the land.

Jamaica

Jamaica, an island nation in the West Indies, is the third largest island in the Caribbean Sea. It lies about 480 miles (772 kilometers) south of Florida, in the tropics. More than 90 per cent of Jamaica's 2-1/2 million people have black African ancestry. Some of these Jamaicans are *Afro-Europeans*—of both African and European ancestry. The country's minority groups include Chinese, East Indians, and other Asians; Europeans; and Syrians.

Modern life

Most Jamaican professional and business people are Europeans or Afro-Europeans. Many Chinese and Syrians are shopkeepers. Large numbers of blacks and Asians work as farm laborers.

About 40 per cent of the people live in rural areas, but many have moved to the cities since the early 1960's. Jamaica's official language is English, but the people actually speak a *dialect* (local form) of English that differs from the language spoken by American or British people.

Around 65 per cent of Jamaicans are Christians. About 50,000 black Jamaicans belong to Ras Tafari, a religious and political movement. Rastafarians, as members are called, consider former Emperor Haile Selassie I of Ethiopia a god. Rastafarians have also adopted many of the beliefs of Marcus Garvey, a Jamaican who died in 1940. Garvey preached that all blacks should consider Africa their home and live there. Rastafarians consider themselves Africans, not Jamaicans. Some believe blacks are a superior race.

History

Arawak Indians were living in what is now Jamaica when Christopher Columbus landed there in 1494. Columbus claimed the land for Spain. The Spaniards enslaved the Arawak people, and almost all the Indians were killed by diseases or overwork. The Spaniards then brought Africans to Jamaica to work as slaves. Spain did not try to settle or develop the island, but used Jamaica mainly as a supply base.

The British invaded Jamaica in 1655 and controlled most of the island within five years. However, they had to keep fighting escaped African slaves called *Maroons,* who had fled to the rugged hill region of Jamaica

called the Cockpit Country when the British arrived. During the 1670's, British pirates used Jamaica as a base to attack Spanish ports and ships in the Caribbean.

Jamaica prospered during the 1700's. Sugar cane became the major crop. The island also ranked as the most important slave market in the Western Hemisphere. In 1738, the British finally signed a peace treaty with the Maroons, on the former slaves' terms. And 100 years later, the British Parliament freed all the slaves on the island. Most of the freed slaves became independent farmers. The sugar industry was hurt because the plantation owners lost thousands of slave laborers.

In 1865, relations between the planters and their workers had deteriorated so much that the workers revolted. The Morant Bay Rebellion, as the revolt was called, was led by Paul Bogle, a Baptist deacon. British troops put down the uprising. The British government then made Jamaica a crown colony, governed entirely by the United Kingdom. They eliminated the House of Assembly that Jamaicans previously had elected to help the British rule the island.

During the 1930's, Jamaican labor leaders urged the United Kingdom to grant Jamaicans more political power. In 1944, a new Constitution gave Jamaicans some self-government. In 1962, Jamaica achieved full independence.

730

Waterfall climbing is a sport unique to Jamaica. *Island of springs* was the original Arawak Indian name for the island.

Balancing bananas on her head, a Jamaican woman helps bring in the harvest, *far left.* About 25 per cent of Jamaica's workers, especially those of African and Asian ancestry, are farmworkers.

Cricket fans are guaranteed an unobstructed view of the game from high atop fence posts, *bottom left.* Cleverly placed coconuts add comfort to their enjoyment of Jamaica's national sport.

A cluster of shacks on the outskirts of Montego Bay, *below,* reflects the continuing movement of poor rural people to the towns.

Jamaica Today

The Arawak Indians, who were the first people to live in Jamaica, called the island *Xaymaca*, which meant *island of springs*. Today, Jamaica's beautiful beaches, soaring mountains, and pleasant climate attract more than 700,000 tourists every year.

The land and climate

Jamaica's many mountains include the Mocho Mountains, which rise in the center of the island, and the Blue Mountains, which rise in the east. Swift rivers flow north and south from the mountains, and the island also has many springs, streams, and waterfalls.

Limestone formations are found in the northwest area known as Cockpit Country. This region was named for the many deep depressions called *cockpits*. Low plains along the coasts are lined with many beautiful sandy beaches.

Jamaica has a tropical climate, but the heat and humidity are moderated by ocean breezes along the coasts and by altitude in the mountains. Temperatures in mountain areas sometimes drop to 40° F. (4° C).

The economy

Thanks to Jamaica's tropical beauty and pleasant climate, the tourist industry is one of

Red-stained water, caused by the mining of bauxite, mars the natural beauty of Jamaica, *below*. Mining brings Jamaica needed income, but the government must ensure that tourist areas are kept free of such pollution.

The island of Jamaica is the third largest Caribbean island. Only Cuba and Hispaniola are larger. Kingston is Jamaica's capital, largest city, and chief port.

FACT BOX

COUNTRY

Official name: Jamaica
Capital: Kingston
Terrain: Mostly mountains, with narrow, discontinuous coastal plain
Area: 4,243 sq. mi. (10,990 km²)
Climate: Tropical; hot, humid; temperate interior

Main rivers: Great, Black, Minho, Cobre, Yallahs
Highest elevation: Blue Mountain Peak, 7,402 ft. (2,256 m)
Lowest elevation: Caribbean Sea, sea level

GOVERNMENT

Form of government: Constitutional parliamentary democracy
Head of state: British monarch, represented by governor general
Head of government: Prime minister
Administrative areas: 14 parishes

Legislature: Parliament consisting of the Senate with 21 members and the House of Representatives with 60 members serving five-year terms
Court system: Supreme Court, Court of Appeal
Armed forces: 2,830 troops

PEOPLE

Estimated 2002 population: 2,628,000
Population growth: 0.46%
Population density: 619 persons per sq. mi. (239 per km²)
Population distribution: 50% urban, 50% rural
Life expectancy in years:
Male: 73
Female: 77
Doctors per 1,000 people: 1.4
Percentage of age-appropriate population enrolled in the following educational levels:
Primary: 98
Secondary: 90
Further: 9

the country's leading economic activities. Tourist centers include Kingston, Montego Bay, Negril, and Ocho Rios.

Jamaica is the world's third largest producer of bauxite, and mining provides much of the nation's income. Plants near some of the bauxite mines remove a mineral compound called *alumina* from the ore—the first step in the production of aluminum. Jamaicans also mine gypsum, which is used in making plasterboard and other construction materials.

Although agriculture employs about a fourth of all Jamaican workers, the farms do not produce enough food for the country, so Jamaica imports much of its food supply. Sugar cane is the nation's most important crop. Other farm products include allspice, bananas, cacao, citrus fruits, coconuts, coffee, milk, and poultry.

Jamaica manufactures cement, chemicals, cigars, cloth, fertilizer, footwear, machinery, petroleum products, and tires. Molasses and rum are produced from the sugar cane grown on the island.

The government

Jamaica was a British colony for about 300 years. Today, the country is an independent nation within the Commonwealth of Nations. The head of state is the monarch of the United Kingdom, represented by a governor general. The governor general has little power, however. The country's chief executive is the prime minister—the leader of the majority party in Parliament, Jamaica's legislature. The Jamaican Parliament consists of a House of Representatives and a Senate. The 60 representatives are elected by the people to five-year terms. The 21 senators are appointed by the governor general, 13 of them on the advice of the prime minister and 8 on the advice of the leader of the minority, or opposition, party in Parliament. The largest political parties are the Jamaican Labor Party (JLP) and the People's National Party (PNP).

Jamaica became independent in 1962. Since then, the country has faced many problems, including poverty, unemployment, and inflation. In the 1970's, Michael Manley of the PNP became prime minister. He tried to solve the country's economic problems by adopting socialist policies. He also began seeking relations with leftist governments, a policy that worried Jamaica's traditional Western allies, such as the United Kingdom and the United States. Jamaica's tourist industry suffered.

Edward Seaga of the JLP became prime minister in 1980 and adopted policies designed to help private business and promote good relations with Western nations. In 1989, Manley became prime minister again, but stated he would follow more moderate economic and foreign policies. Manley resigned in 1992. Percival J. Patterson was elected head of the PNP and became prime minister.

Languages spoken:
English
Creole

Religions:
Protestant 61%
Roman Catholic 4%
spiritual cults

TECHNOLOGY

Radios per 1,000 people:
784

Televisions per 1,000 people: 194

Computers per 1,000 people: 46.6

ECONOMY

Currency: Jamaican dollar

Gross national income (GNI) in 2000: $6.9 billion U.S.

Real annual growth rate (1999–2000): 0.8%

GNI per capita (2000): $2,610 U.S.

Balance of payments (2000): -$275 million U.S.

Goods exported: Alumina, bauxite, sugar, bananas, rum

Goods imported: Machinery and transport equipment, construction materials, fuel, food, chemicals, fertilizers

Trading partners: United States, European Union, Canada, Caricom countries

Reggae Music

Jamaica has long been known for its tropical beauty and pleasing climate. It has also become known as the birthplace of a popular kind of music.

Many of the Caribbean islands have their own special kind of music. In Cuba, the music is salsa; in Trinidad, it is calypso; and in Jamaica, it is reggae. For many Jamaicans, however, reggae is more than a local music; it is closely connected with politics, religion, and pride in their African roots.

Reggae arose in Jamaica from many different kinds of music. It grew out of Jamaican folk music as well as ska and rock-steady, which, in turn, grew out of American rhythm and blues and soul. Its regular, steady rhythm can be traced to African music. The basic instrumental arrangement of electric bass, rhythm guitar, keyboard, drum kit, and horns is accompanied by a distinctive kind of singing.

Reggae began to develop in the 1960's as young musicians in Kingston sought to reflect the tensions of the times in their music. Jamaica had achieved independence, but poverty and unemployment were widespread. Many reggae songs demanded social and political changes. Local Jamaican bands made the music popular on the island, and politicians came to use reggae music and musicians in their struggle for supporters.

Reggae gained international popularity largely through a musician and singer named Bob Marley. Born in rural Jamaica in 1945, Marley died in 1981 of cancer at 36, but he had already become a legendary figure both in Jamaica and elsewhere.

Marley began singing and writing music in his teens. His group, the Wailers, became one of the most popular reggae bands in Jamaica in the early 1960's and had become famous outside Jamaica by 1973. Many of his songs express the beliefs of Ras Tafari, a religious and political movement to which Marley belonged.

Admirers are welcomed at both Marley's recording studio and his former home, which opened as a museum dedicated to him in 1986. A mural depicting his life is included in a tour of the museum. A heart in the center of the mural represents the heart of the Jamaican people and the love they have for Marley.

Ziggy Marley and the Melody Makers is led by Bob Marley's son, *below.*

The Reggae Sunsplash festival was held every year in Montego Bay, Jamaica, and featured a variety of reggae artists, including Lieutenant Stitchie seen here, *below at right.* Sunsplash was once the best-known reggae festival.

The Bob Marley Museum, *right,* displays the colors of the Rastafarians and the Ethiopian flag. The museum attracts fans of the reggae star.

Jamaica was the birth-place of reggae. Other West Indian islanders have produced their own kinds of music.

Kingston's Trench Town, *below,* is the birthplace of many reggae bands.

Robert Nesta Marley, *above,* known as Bob Marley, made reggae music popular around the world. The music became a vehicle for black pride. In recognition of his achievement, Jamaica honored him with the Order of Merit award in 1981.

Located in Kingston, the Bob Marley Museum is easily recognized because red, yellow, and green stripes mark the entrance, and the Ethiopian flag flies outside. Red, yellow, and green are the colors of Ras Tafari.

The Rastafarian movement began in Jamaica in the 1920's. It was named for Ras (Duke) Tafari Makonnen, who took the title Haile Selassie I when he became emperor of Ethiopia in 1930. Several Jamaican religious sects had predicted that a black emperor would be a savior of black people around the world. Rastafarians consider Selassie to be God. Rastafarian practices include avoiding foods considered impure and not cutting the hair—leading to the braided hair style called *dreadlocks.*

Rastafarians also believe in the teachings of Marcus Garvey, a black leader who started a "back to Africa" movement. Garvey was born in Jamaica and started his movement there, then moved to the United States. Garvey believed that blacks would never receive justice in countries where most of the people were white. He preached that blacks should consider Africa their homeland.

Reggae music arose from the ethnic pride of the Rastafarian movement. The reggae music of Bob Marley overcame racial and cultural barriers to become popular around the world.

With his popularity, Marley influenced many musicians. Not only have his songs been covered with much success by other performers—like Eric Clapton's version of "I Shot the Sheriff," a hit in the United States—his children carried on his music. His son David "Ziggy" Marley fronts the band Ziggy Marley & the Melody Makers, which includes several of his siblings. Marley's children also perform in other groups including the Ghetto Youths Crew and The Marley Girls.

The Jamaican tourist spot of Montego Bay held a concert festival called Reggae Sunsplash for many years. Currently, Jamaica has a reggae festival every summer called Sumfest. Reggae music festivals are also popular in many other countries, and the reggae sound is heard in nightspots throughout Jamaica.

Japan

Japan, a beautiful island country in the North Pacific Ocean, lies off the northeast coast of mainland Asia and faces southeastern Russia, the Korean Peninsula, and China. Japan's four major islands—Hokkaido, Honshu, Kyushu, and Shikoku—and thousands of smaller ones form a curved landmass that extends for about 1,200 miles (1,900 kilometers). More than 126 million people are crowded onto these islands, making Japan—with 867 people per square mile (335 per square kilometer)—one of the world's most densely populated countries.

The Japanese call their country Nippon or Nihon, which means *source of the sun.* The name *Japan* may derive from *Zipangu,* a name given to the country by the Italian trader Marco Polo who heard about the Japanese Islands while traveling throughout China in the 1200's.

Framed by cherry blossoms, majestic Mount Fuji symbolizes the great natural beauty of Japan. Because mountains and hills take up most of the land, the great majority of the Japanese people live on the narrow coastal plains. These plains have much of the country's best farmland and most of its largest cities, which are centers of commerce, culture, and industry. Tokyo, the country's largest city, is also its capital. More than 75 per cent of the people live in urban areas.

Early Japan was greatly influenced by China and borrowed heavily from Chinese art, government, language, religion, and technology. Japan closed its doors to Western influence in the early 1600's but renewed relations with the West in the mid-1800's. By the early 1900's, Japan had become a leading industrial and military power. However, the country suffered a devastating defeat in World War II (1939–1945). By 1945 many Japanese cities lay in ruin, industries were shattered, and Allied forces occupied the country.

The Japanese worked hard to rebuild their country, concentrating their efforts on economic development. By the late 1960's, Japan had become a great industrial nation. Today, the country is one of the world's economic giants with a total economic output exceeded only by that of the United States.

While Japan's big cities look much like those of Western nations and the Japanese people enjoy a high standard of living, life in Japan reflects the culture of both the East and the West. For example, baseball games and exhibitions of traditional sumo wrestling are the nation's favorite sporting events.

Japan Today

CHINA

NORTH KOREA

SOUTH KOREA

CHINA

Sea of Japan

East China Sea

PACIFIC OCEAN

Philippine Sea

Okhotsk

Rebun I.
Rishiri I.
Wakkanai
Teshio
Abashiri
Hokkaido I.
Rumoi
Ishikari
Kitami
Kamui Cape
Asahigawa
Iwamizawa
Nen
Otaru
Yubari
Kushiro
Sapporo
Obihiro
Tomakomai
Muroran
Uchiura Bay
Erimo Cape
Okujiri I.
Hakodate
Tsugaru Strait
Shiriya Cape
Seikan Tunnel
Mutsu Bay
Aomori
Hachinohe
Hirosaki
Noshiro
Nyudo Cape
Morioka
Akita
Kamaishi
Sakata
Tsuruoka
Izumi
Ishinomaki
Yamagata
Sendai
Sado I.
Yonezawa
Niigata
Fukushima
Nagaoka
Aizuwakamatsu
Koriyama
Noto Peninsula
Toyama Bay
Takaoka
Nagano
Maebashi
Hitachi
Kanazawa
Toyama
Utsunomiya
Komatsu
Takasaki
Mito
Fukui
Matsumoto
Omiya
Wakasa Bay
Kofu
Urawa
Funabashi
Maizuru
TOKYO
Chiba
Tottori
Nagoya
Kawasaki
Matsue
Gifu
Seto
Mt. Fuji
Yokohama
Oki Is.
Kyoto
Yokkaichi
12,388 ft (3,776 m)
Yokosuka
Dago I.
Chugoku Mts.
Himeji
Toyohashi
Shizuoka
Sagami Bay
Okayama
Tsu
Himeji
Kobe
Hamamatsu
Hiroshima
Biwa Lake
Osaka
Suruga Bay
Yamaguchi
Kure
Harima Sea
Sakai
Ise Bay
Miyake I.
Izu Islands
Shimonoseki
Takamatsu
Wakayama
Honshu I.
Ube
Matsuyama
Kitakyushu
Tokushima
Kii Peninsula
Suo Sea
Kochi
Korea Strait
Iki I.
Iyo Sea
Shiono Cape
Hachijo I.
Tsushima
Saga
Fukuoka
Sasebo
Omuta
Oita
Uwajima
Shikoku I.
Hirado I.
Kumamoto
Bungo Channel
Goto Is.
Nagasaki
Nobeoka
Amakusa Is.
Kyushu I.
Koshiki Is.
Miyazaki
Aoga I.
Kagoshima
Miyakonojo
Satsuma Peninsula
Osumi Peninsula
Sumisu I.
Osumi Is.
Tanega I.
Tori I.
Yaku I.
Tokara Is.
Amami I.
Tokuno I.
Ryukyu Islands
Okinawa I.
Senkaku I.
Naha
Iriomote I.
Miyako I.
Ishigaki I.
Daito Islands

N

0 km   250   500
0 miles   250

FACT BOX

JAPAN

Japan became the world's wealthiest nation in 1987. However, many of its allies believe that Japan should spend more money on defense and take a more active role in regional defense arrangements.

Emperor Akihito, *above,* took the throne in 1989, promising to fulfill his role as a symbol of the Japanese people's unity—the 1947 Constitution grants the emperor only ceremonial duties. The Japanese government chose the name *Heisei,* which means *peace and concord,* for the new emperor's reign.

In the early 1990's, Japan had the highest rate of economic growth among the leading industrialized nations. Although Japan had been learning from the West, adopting its ideas and technology since the mid-1800's, in the 1980's Western nations began to look to Japan for ways to improve their own economies.

The 1980's were also marked by tension between Japan and its trading partners over economic issues. The countries that traded with Japan complained that competition from Japanese goods was damaging some of their domestic industries, and they criticized Japan for setting up trade barriers that limited its imports. In 1981, to improve trade relations, Japan agreed to set limits on its exports of automobiles to Canada, the United States, and West Germany and also pledged to remove some restrictions on its imports.

Within Japan, the Japanese people were distressed by the death of Emperor Hirohito in January 1989. Hirohito had reigned since 1926. His son, Akihito, assumed the throne upon the emperor's death and pledged to protect the Japanese Constitution. Drawn up by the Allied occupation forces, the 1947 Constitution transferred all political power from the emperor to the Japanese people, took away Japan's right to wage war, and abolished its armed forces. However, Japan now maintains air, ground, and sea forces for purposes of self-defense.

The Constitution guaranteed the people many human rights and provided for three branches of government—legislative, executive, and judicial. The legislative branch is composed of a two-house parliament called the *Diet.* The Diet makes Japan's laws. In addition, its members choose the prime minister, the country's chief executive.

Japan has several political parties, but the Liberal-Democratic Party (LDP), a conservative party, controlled the government from 1955 to 1993. In 1993, the LDP split and lost power. In 1994, Japan had three different prime ministers: Morihiro Hosokawa, who headed a coalition government but resigned amid accusations of bribery; former LDP member Tsutomo Hata, who was prime minister for only two months; and Socialist Tomiichi Murayama.

On Jan. 17, 1995, an earthquake measuring 7.2 on the Richter scale shook the city of Kobe. About 5,000 people died, and damage was estimated at between $70 billion and $100 billion.

COUNTRY

Official name: Japan
Capital: Tokyo
Terrain: Mostly rugged and mountainous
Area: 145,883 sq. mi. (377,835 km²)
Climate: Varies from tropical in south to cool temperate in north

Main rivers: Ishikari, Shinano, Tone
Highest elevation: Fujiyama, 12,388 ft. (3,776 m)
Lowest elevation: Hachiro-gata, 13 ft. (4 m) below sea level

GOVERNMENT

Form of government: Constitutional monarchy
Head of state: Emperor
Head of government: Prime minister
Administrative areas: 47 prefectures
Legislature: Kokkai (Diet) consisting of the Sangi-in (House of Councillors) with

252 members serving six-year terms and Shugi-in (House of Representatives) with 500 members serving four-year terms
Court system: Supreme Court
Armed forces: 236,300 troops

PEOPLE

Estimated 2002 population: 127,018,000
Population growth: 0.18%
Population density: 871 persons per sq. mi. (336 per km²)
Population distribution: 78% urban, 22% rural
Life expectancy in years:
Male: 78
Female: 84
Doctors per 1,000 people: 1.9
Percentage of age-appropriate population enrolled in the following educational levels:
Primary: 102*
Secondary: 102*
Further: 44

Language spoken: Japanese
Religions: Observing both Shinto and Buddhist 84% Christian 1%

*Enrollment ratios compare the number of students enrolled to the population which, by age, should be enrolled. A ratio higher than 100 indicates that students older or younger than the typical age range are also enrolled.

TECHNOLOGY

Radios per 1,000 people: 956
Televisions per 1,000 people: 725
Computers per 1,000 people: 315.2

ECONOMY

Currency: Yen
Gross national income (GNI) in 2000: $4,519.1 billion U.S.
Real annual growth rate (1999–2000): -2.4%
GNI per capita (2000): $35,620 U.S.
Balance of payments (2000): $116,883 million U.S.
Goods exported: Motor vehicles, semiconductors, office machinery, chemicals
Goods imported: Fuels, foodstuffs, chemicals, textiles, office machinery
Trading partners: United States, China, Taiwan, South Korea

Environment

The Japanese islands are formed by the upper part of a great mountain range that rises from the floor of the North Pacific Ocean. Mountains and hills cover about 70 per cent of Japan's land area, and dense forests cover about 68 per cent of the mountainsides, adding to the beauty of the Japanese islands.

Japan lies on an extremely unstable part of the earth's crust and is therefore subject to frequent earthquakes and volcanic activity. Most of the roughly 1,500 earthquakes that Japan experiences every year are minor, but severe quakes occur every few years. In addition, undersea earthquakes sometimes cause huge, destructive tidal waves, called *tsunamis,* along Japan's Pacific coast. The Japanese islands also have more than 150 volcanoes, 60 of which are active.

Japan is composed of four main islands and thousands of smaller islands and islets. In order of size, the main islands are Honshu, Hokkaido, Kyushu, and Shikoku.

The main islands

Honshu, Japan's largest island, is home to about 80 per cent of the Japanese people. Three mountain ranges cross northern Honshu, and two plains that support some farming lie to the east and west of these mountains. The Japanese Alps, Japan's highest mountains, rise in central Honshu. A chain of volcanoes that cuts across the center of the island includes Mount Fuji, an inactive volcano and the country's tallest and most famous peak. Farther east lies the Kanto Plain, the country's largest lowland area as well as an important center of agriculture and industry. Tokyo stands on the Kanto Plain. Two other major agricultural and industrial lowlands lie south and west of the Kanto region. Mountains cover most of southwestern Honshu.

Hokkaido, the northernmost of Japan's four major islands, is the country's second largest island but has only about 5 per cent of the total population. The Ishikari Plain in west-central Hokkaido is the largest lowland and the chief farming region on the island. Smaller plains border the island's east coast, and much of the rest of Hokkaido consists of forested mountains

and hills. The island is a popular area for winter sports.

Kyushu, the southernmost of the main islands, has the second-highest population after Honshu, with about 11 per cent of Japan's population. Most of the people live in the heavily industrialized lowlands of northwestern Kyushu and in the major farming districts along the west coast. A chain of steep, heavily wooded mountains runs down the center of the island. The volcanic regions in the northeastern and southern sections of Kyushu contain only small patches of farmland.

Shikoku, the smallest of the main Japanese islands, lies off the coast of southwestern Honshu and has only about 3 per cent of Japan's total population. Most of the people on this largely mountainous island live in northern Shikoku, where the land slopes down to the Inland Sea. Farmers grow rice and a variety of fruits on the fertile land along the Inland Sea, and some copper mining is carried out. Hundreds of hilly, wooded islands dot this body of wa-

Mount Aso, a volcano on the island of Kyushu, is surrounded by lush green fields and scattered farming settlements. Lakes have formed in the craters of some extinct volcanoes.

Tiny offshore islets, as well as large mountainous islands, make up Japan's landscape, *below.* Most of Japan's people live near the coasts. The mountainous interiors of the islands are thinly populated.

ter. A narrow border along Shikoku's southern coast also supports some farming.

Climate

Seasonal monsoons and two Pacific Ocean currents—the Japan Current and the Oyashio Current—affect Japan's climate. Hokkaido and northern Honshu, influenced by the cold Oyashio Current and by monsoons from the northwest in winter, have cool summers and severe, snowy winters. The southern and eastern areas of the country are warmed by the Japan Current and by monsoons from the southeast in summer. These areas have mild winters and warm, humid summers.

All areas of the country—except eastern Hokkaido—receive at least 40 inches (100 centimeters) of rain yearly, mainly in the summer and early fall. In addition, several typhoons hit Japan each year, chiefly in late summer and early fall. These storms often do great damage to houses and crops.

Nikko National Park, *left,* on Honshu has a magnificent scenic landscape with high peaks, rocky gorges, and thundering mountain waterfalls. Many vacationers visit the park to enjoy some of Japan's most spectacular scenery.

The hourly progress of a tsunami produced by an earthquake that originated south of Alaska is shown on the map at the right. *Tsunami* is a Japanese word meaning *harbor wave.* These tidal waves generally occur along fault lines in the Pacific Ocean. A tsunami can travel 500 to 600 miles (800 to 970 kilometers) per hour in open waters. As the tsunami approaches shallower water near shore, it may form a gigantic wall of water more than 100 feet (30 meters) high.

During an undersea earthquake, the shifting rocks of the sea floor create a wave in the water above, *right.* The wave may begin small in deep ocean water, but it can develop into a towering tsunami as it nears the shore.

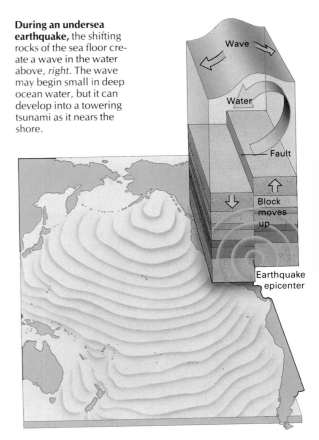

741

Island Groups

The Bonin, Izu, Ryukyu, and Volcano island chains are among the thousands of smaller islands that make up Japan's land area. Although Japan also claims several islands in the southern part of the Kurils, Japanese control of these islands was taken away at the end of World War II. The southern Kuril Islands, which lie off the northeast coast of Hokkaido Island, are referred to by the Japanese as the Northern Territories.

The Izu Islands are a group of volcanic islands that stretch southward off the coast of Honshu. Most of the people on these islands work on farms or fish in coastal waters. Farther south lie the three islands that comprise the sparsely populated Volcano Islands. Iwo Jima, the middle island, was captured from the Japanese by American forces during World War II. The United States controlled the island until 1968, when it was returned to Japan.

The Bonin Islands lie between the Izu and Volcano island chains, about 600 miles (970 kilometers) south of Tokyo. The 97 volcanic islands in the Bonin Island chain have a total area of about 41 square miles (106 square kilometers). The rocky, rugged land on the islands is covered with scrubby trees and tall grass. About 1,900 people live on the islands. They raise cacao, cattle, fruits, sugar cane, and vegetables and make coral ornaments.

Japan claimed the Bonins in 1875 and named them *Ogasawara-gunto*. U.S. forces attacked the Japanese stationed on the Bonins during World War II. The islands were placed under U.S. control after the war, but they were returned to Japan in 1968. Today, visitors enjoy the islands' warm, mild climate and white sand beaches.

The Ryukyu Islands are a group of more than 100 islands that extend from the southern tip of Kyushu Island to Taiwan. They have a land area of 1,205 square miles (3,120 square kilometers) and a population of about 1 million. Some of the islands, particularly those containing active volcanoes, are uninhabited.

Most of the Ryukyuans are farmers. They grow rice, but their main food crop is sweet potatoes. They also raise and export sugar cane and pineapples. Fishing is another important activity for the Ryukyuans, bringing food and income. The people speak a language similar to Japanese, and their religious beliefs have been heavily influenced by both Japan and China.

Okinawa, the largest and most important of the Ryukyu Islands, covers an area of 554 square miles (1,434 square kilometers) and has about 1 million people. Naha, the capital and largest city of the islands, is on Okinawa.

Like the rest of the islands in the Ryukyu chain, Okinawa is largely mountainous and has a warm, wet climate. Farmers on Okinawa raise much the same crops

The beautiful beaches and warm weather of the Bonin Islands, *above*, attract many tourists, mainly from other parts of Japan. The Japanese government declared the Bonins a national park region in 1972 to protect the islands' natural beauty.

Sugar cane is one of the chief crops raised on Okinawa as well as on the other islands, *above left*, of the Ryukyu chain. Although agriculture is important to the economy of the islands, fishing and tourism also produce income.

The modern town hall in Nago on Okinawa was constructed after World War II. Most of the island's buildings were destroyed during the war and had to be replaced.

grown throughout the Ryukyus, but the island's economy depends largely on tourism and United States military spending.

Japan and China both claimed Okinawa and the rest of the Ryukyus until 1874, when China signed a treaty recognizing Japanese rule. One of the bloodiest battles of World War II was fought on Okinawa between U.S. and Japanese troops in 1945. More than 50,000 American troops were killed and about 110,000 Japanese died. About 90 per cent of the island's buildings were destroyed during the campaign.

After Japan was defeated, the United States took over the Ryukyus. In 1953, the United States returned the northern islands to Japan, and Okinawa and the southern Ryukyus were returned in 1972. Under an agreement between the United States and Japan, U.S. military bases remain on Okinawa, but nuclear weapons may not be kept on the island without Japan's consent.

Many of Japan's smaller island groups lie south of the four main Japanese islands. The Ryukyu Islands, including Okinawa, extend to the southwest. The Izu, Bonin, and Volcano island chains stretch to the southeast. The map below is an enlarged view of the region indicated in the map at the left.

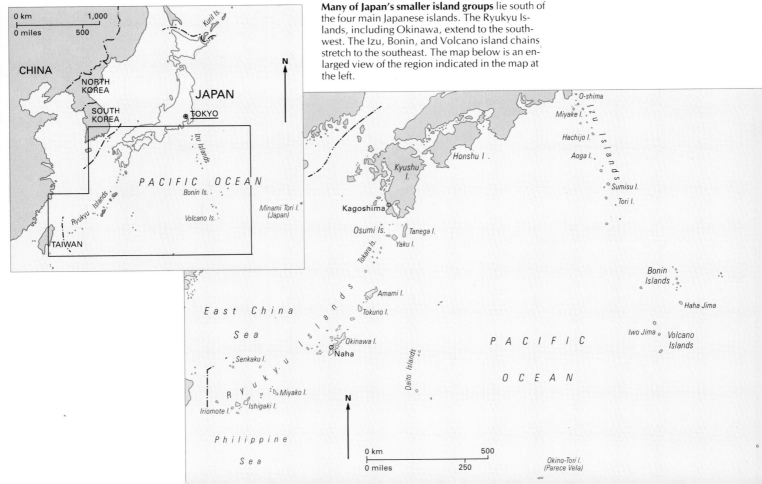

Agriculture

As a result of Japan's massive industrial development after World War II, the total labor force employed in agriculture fell from about 50 per cent before the war to about 9 per cent today. Land reform and modern technology increased agricultural production and greatly reduced the number of workers required. Today, agriculture accounts for only 3 per cent of Japan's *gross domestic product,* the total value of all goods and services produced within a country in one year.

Although only about 15 per cent of Japan's land can be cultivated, Japanese farmers produce 70 per cent of the food needed to feed the nation's people. The farms average only about 2.4 acres (1 hectare), but they produce extremely high yields. Japanese farmers make their land as productive as possible through the use of irrigation, improved seed varieties, and modern agricultural chemicals and machinery. In the country's mountainous regions, farmers grow crops on terraced fields.

Before World War II, many Japanese farmers rented their land and paid their landlords as much as 50 per cent of their harvest. After the war, a land-reform program reduced the holdings of the landlords and enabled the farmers to buy the land they worked. Today, about 90 per cent of Japan's farmers own their farms.

More than 50 per cent of Japan's farmland is used to grow rice, the country's staple food. Rice is grown throughout the Japanese islands, and production exceeds demand.

Japanese farmers work in rice fields on the southern island of Shikoku, *right.* Japan is one of the world's leading rice-producing countries.

Bamboo shoots, widely used in Asian cooking, are grown all over Japan.

Japanese women pick tea leaves on an emerald-green hillside. Because Japan is so mountainous, level farmland is scarce. As a result, many Japanese farmers grow crops on *terraced fields*—that is, level strips of land cut out of the hillsides.

Japanese farmers also raise a wide variety of other crops, including sugar beets, tea, tobacco, and wheat. The leaves of mulberry bushes, which are grown on some hillsides, feed silkworms used in the production of raw silk. Apples, cabbages, citrus fruits, eggplants, pears, potatoes, strawberries, and tomatoes are among the fruits and vegetables grown in Japan. In addition, because the Japanese people have increased their consumption of meat and dairy products, more farmers are raising beef and dairy cattle, chickens, and hogs.

Fish, on the other hand, has provided the chief source of protein in the Japanese diet for centuries. Japan's fishing industry employs only about 1 per cent of the country's total labor force, yet it is the largest in the world. Japan has over 400,000 fishing vessels—more than any other country— and they catch about 13 million short tons (12 million metric tons) of fish yearly.

Japan leads the world in tuna fishing and ranks second to the United States in its catch of salmon. Flatfish, mackerel, pol-lock, and sardines are also important to the nation's fishing industry. In addition, Japanese fishing crews catch large quantities of shellfish and harvest oysters and edible seaweed from ''farms'' in coastal waters.

Japan's fishing fleets operate all over the world, but their activities have been limited since the 1970's, when almost all coastal nations established protected fishing zones extending 200 nautical miles (370 kilometers) off their shores. These protected zones excluded Japanese fleets from some of their valuable fishing grounds and thereby reduced Japan's annual catch. Industrial pollution in the country's own coastal waters has further reduced the catch. As a result of this decline in production, Japan has had to import some seafood to meet its needs.

For many years, Japan was a leading whaling nation. International concern over this endangered species eventually persuaded Japan to limit its whale catch. Finally, in 1988 Japan agreed to join the other countries of the world in imposing a temporary halt on all commercial whaling.

Dockworkers handle a huge tuna catch at a Japanese port. To satisfy the Japanese people's tremendous appetite for seafood, Japan operates the world's largest fishing industry.

Industry

Japan began its industrialization process in 1868 under the Meiji leaders, when the government established a number of experimental industrial projects to serve as models for new industries. American and European experts in many fields were hired to teach Western knowledge and methods to the Japanese people, and by the early 1900's Japan had achieved significant industrial growth and developed some foreign trade.

By the 1920's, Japan's most important industries were under private ownership. Financial institutions and manufacturing, mining, and trading companies were controlled by the *zaibatsu,* huge corporations owned by single families. In the late 1930's and early 1940's, the output of the country's modern industries doubled as Japan prepared for war.

Japan's defeat in World War II, however, severely damaged its economy, and destroyed half of the country's industrial capacity. But the Japanese amazed the world by quickly overcoming the effects of the war. Japan's recovery was aided in large part by financial aid from the United States. The Japanese invested in the latest technology to make their postwar industries highly productive, and labor unions were established. By the late 1960's, Japan was a leading industrial nation.

Manufacturing and service industries

Manufacturing, which employs about 24 per cent of the country's workers, is the single most important economic activity in Japan. It accounts for 28 per cent of the country's gross domestic product.

The production of transportation equipment is the country's most important manufacturing industry. Japan is the world's leading producer of ships and cars, and among the top producers of iron and steel. Japan's thriving chemical industry produces petrochemicals and petrochemical products, such as plastics and synthetic fibers. In addition, Japan is one of the leading producers of ceramics, paper products, raw silk, and textiles.

One of the fastest-growing industries in the nation is the production of machinery, including heavy electrical and nonelectri-

The world's leading ship-builder and carmaker, Japan is also among the top iron and steel producers. However, the country must import nearly all of its petroleum and all of its natural gas to meet the needs of its manufacturing industries.

Passenger cars are Japan's chief exports, *right*. As a result of protest against its trade policies, however, Japan reduced its exports of automobiles to the United States, its main trading partner, in the 1980's.

cal machines, electrical appliances, and electronic equipment. Japan is also noted for its production of precision instruments, such as cameras, clocks and watches.

Service industries, which employ about 56 per cent of the country's workers, are another important economic activity in Japan. Service industries include banks and other financial institutions, government agencies, trade, and transportation and communication. Altogether, these economic activities account for 61 per cent of Japan's gross domestic product.

Foreign trade

Because Japan has limited natural resources, it must import many of the raw materials needed by its manufacturing industries. Japan sells its manufactured goods throughout the world to pay for its imports. Thus, the Japanese economy depends heavily on foreign trade.

Since the mid-1960's, the country has usually maintained a favorable *balance of trade* (the difference between the value of a nation's exports and the value of its imports). However, in the late 1970's and early 1980's, many of Japan's trading partners criticized the country's trade policies. They claimed that while Japan exported large quantities of competitively priced goods to foreign markets, it imposed trade barriers on imports from other countries. These policies were creating unfavorable trade balances for Japan's trading partners, especially the United States.

To maintain good trade relations, Japan has begun to limit some of its exports and to lift obstacles to imports. Nevertheless, some U.S. officials still criticize the serious trade imbalance that remains in Japan's favor.

The Tokyo Stock Exchange is one of the world's leading market places. This financial institution is one of Japan's important service industries.

The bullet-shaped train called the *Shinkansen, below left,* is one of the fastest trains in the world. Japan has a modern, highly efficient transportation system, including highways, railroads, and coastal shipping. All the major cities have extensive local-transit networks that include buses, trains, and subways.

Electronic products, *center,* are among Japan's most valuable exports. High-quality Japanese calculators, computers, phonographs, radios, tape recorders, television sets, and videocassette recorders are sold throughout the world.

747

History

Scientists believe that communities of hunters and gatherers lived on the Japanese islands at least 6,500 years ago. By 200 B.C., farming was developed, and the people lived in villages.

From the A.D. 200's, Japan was controlled by warring clans. The Yamato clan established its authority in the 400's. Under the Yamato, Japan borrowed many ideas and technologies from China. Under the Taika Reform program in 646, Japan also adopted many features of China's centralized imperial government.

In 794, the Japanese capital was established at Heian, later called Kyoto. The Fujiwaras, a powerful noble family, gained control of the emperor and his court in 858 and ruled Japan for about 300 years. During this time, the emperors officially reigned, but they lost all real power.

Powerful bands of warriors called *samurai* fought for control of the imperial court during the 1000's. The Minamoto family headed one of these bands. In 1192, the emperor gave Yoritomo Minamoto the title *shogun* (general), and Yoritomo's military government in Kamakura became known as the *shogunate*. The shoguns controlled Japan until 1867, but always in the name of the emperor.

In 1543, Portuguese sailors became the first Europeans to reach Japan. Eventually, Roman Catholic missionaries from Portugal and Spain converted many Japanese to Christianity. However, Ieyasu Tokugawa, who established his shogunate in 1603, feared that the Christian missionaries might bring European armies with them to conquer Japan. As a result, the Tokugawa government cut ties with other nations during the 1630's, and Japan was isolated from the rest of the world for more than 200 years.

Between 1853 and 1854, Commodore Matthew C. Perry of the United States sailed warships into the bay at Edo (Tokyo)

The Toshogu Shrine, located north of Tokyo in Nikko National Park, holds the tomb of Ieyasu Tokugawa, the shogun who ruled Japan from 1603 to 1616. Ieyasu re-established a strong central government in Japan.

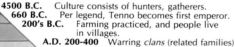

4500 B.C. Culture consists of hunters, gatherers.
660 B.C. Per legend, Tenno becomes first emperor.
200's B.C. Farming practiced, and people live in villages.
A.D. 200-400 Warring *clans* (related families) control Japan.
A.D. 400's Japanese adopt Chinese writing system and calendar.
646 Taika Reform sets up central government controlled by emperor.
858 Fujiwara family gains control of imperial court.
1192 Yoritomo becomes first shogun.
1543 Portuguese sailors become first Europeans to reach Japan.
1603 Tokugawa shogunate founded.
1630's Japan begins isolation.

1853 and 1854 Commodore Perry of the United States opens two Japanese ports.
1867 Tokugawa shogunate overthrown. Emperor regains traditional powers.
1868 Mutsuhito announces Japan's intention to become a modern industrial nation.
1894-1895 Japan defeats China.

1904-1905 Japan is established as a world power following victory in Russo-Japanese War.
1914 Japan enters World War I on the Allied side.
1923 Earthquake destroys much of Tokyo and Yokohama.
1931 Japan seizes Manchuria.
1937 Japan goes to war against China.
1941 Japan attacks U.S. bases at Pearl Harbor.
1945 Japan surrenders after atomic bomb attacks on Hiroshima and Nagasaki.
1950's Akira Kurosawa becomes first Japanese motion-picture director to gain international fame.
1951 Japan signs security treaty with United States.
1952 Allied occupation of Japan ends.
1989 Emperor Hirohito dies. Succeeded by Crown Prince Akihito.
1994 Tomiichi Murayama becomes Japan's first socialist prime minister since 1948.

Ieyasu Tokugawa closed Japan to the outside world in the 1630's.

Emperor Mutsuhito led Japan through industrialization in the late 1800's.

Akira Kurosawa gained international fame as a film director.

Persecution of Christians and foreigners marked the beginning of Japan's isolation in the 1600's. The Tokugawa government believed that domestic order depended on preventing all contact with the outside world. Japanese were not allowed to leave the country, and some foreign sailors who had been shipwrecked on Japan's shores were killed.

and presented U.S. demands to Japan. Later, the Tokugawa government signed a treaty granting the United States trading rights in two Japanese ports. Other European nations soon signed similar treaties with Japan.

In 1867, the Tokugawa shogunate was overthrown, and the emperor regained his traditional power. In 1868, Emperor Mutsuhito moved Japan's capital from Kyoto to Tokyo and announced the return of imperial rule. Emperor Mutsuhito adopted the title *Meiji,* meaning *enlightened rule.* During the Meiji period, from 1867 to 1912, Japan developed into a modern industrial and military power.

During this time, Japan expanded its territory as a result of three wars: the first Chinese-Japanese War (1894-1895), the Russo-Japanese War (1904-1905), and World War I (1914-1918). By 1918, the Japanese Empire controlled Taiwan, Korea, the southernmost tip of Manchuria, and several Pacific Islands.

In 1931, Japan seized all of Manchuria, and open warfare between Japan and China began in 1937. Meanwhile, Japan had signed anti-Communist pacts with Nazi Germany and Fascist Italy. At the outbreak of World War II (1939-1945) in Europe, Japan allied itself with Germany and Italy. In 1941, Japan attacked U.S. bases at Pearl Harbor in Hawaii in order to bring the United States into the war as well.

The war was brought to an end in 1945 after the United States dropped atomic bombs on the Japanese cities of Hiroshima and Nagasaki in August. Japan suffered a crushing defeat in the war—Japanese casualties numbered in the millions, and much of the country had been destroyed. Japan lost all its territories, keeping only its four main islands and small islands nearby. Allied military forces occupied Japan until 1952, after a security treaty had been signed with the United States in 1951.

The wedding of Prince Aya, Emperor Akihito's son, to Kiko Kawashima took place in 1990, *above.* Like his father, Prince Aya broke with tradition and married a woman who did not belong to the Japanese nobility.

The expansion of the Japanese empire—which had begun in the late 1800's—accelerated with conquests of the East Asian mainland in the 1930's. The empire continued to expand during World War II when Japan took much of Southeast Asia and many Pacific islands.

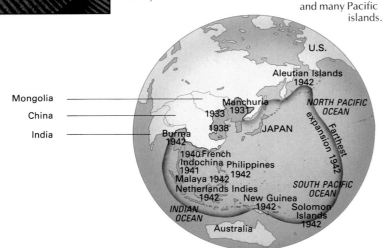

The Samurai

By about 858 A.D., aristocratic families began establishing great private estates in the Japanese countryside. By the 1000's, these estates had become increasingly independent. The lords who controlled the estates were called *daimyo*. They hired bands of warriors to protect their lands and the peasants who worked them. These warriors became known as *samurai*.

The samurai developed from bands of warriors called *bushidan,* who were kinsmen of the estate lords they served. The bushidan consisted of loosely formed armies that usually dissolved after fighting a particular battle. After the rise of the feudal system, however, the bushidan became much more tightly organized. They were composed not only of members of the daimyo's family but also included men who were not related by blood. By the time of the Tokugawa shogunate in 1603, the term *samurai* referred to the entire military class, including the samurai warriors, the daimyos, and the shogun. About 5 per cent of Japan's population belonged to this elite class.

The Bushido code

A code of unquestioning obedience and loyalty, called *Bushido,* bound the samurai warriors to their lords. According to this code, a samurai was expected to lay down his life for his daimyo. And, prizing his honor above his life, a samurai atoned for dishonor by committing *hara-kiri,* or ceremonial suicide.

In service to their daimyo, the samurai warriors wore two swords—the *katana,* a long sword with a slightly curved blade, and the *wakizashi,* a shorter sword. The samurai also carried a weapon called a *naginata,* a blade mounted on a long pole. Samurai warriors usually engaged in hand-to-hand combat using their swords or naginata.

In the late 1500's, Hideyoshi, a great warrior who controlled Japan at that time, ruled that only the samurai were allowed to carry and use weapons. Until this time, soldiering had been combined with farm-

ing and other professions, and even Buddhist temples contained their own arsenals. Hideyoshi carried out a series of raids to disarm farmers and Buddhist monks. This move made the military into a distinct class.

The samurai were graded in military ranks, each with an appropriate income, paid in rice. A samurai retained his status as long as he remained in the service of his daimyo, but once his lord was overthrown, the samurai became a *ronin,* a warrior without a lord. By the 1600's, large numbers of these lordless warriors were working in nonmilitary occupations, and by 1700 most of the ronin had taken administrative posts and had lost both the will and the skill to fight effectively.

Modern samurai

The restoration of imperial rule in 1868 and the rapid modernization of Japan marked the end of the samurai. After the Japanese abolished feudalism in 1871, the samurai lost their privileges as a distinct class. Also, the government ruled in 1876

Yoritomo, shown riding a black horse at the far left of the painting at the left, became the first shogun in 1192. Yoritomo headed the Minamoto family, the strongest family in Japan at that time. A powerful band of samurai warriors protected the shogun.

Samurai warriors, *above,* wore distinctive headdresses and magnificent protective armor that identified the family the warrior served.

The Jidai Matsuri, which means *Festival of the Ages,* is celebrated in Kyoto every October, *left.* More than 2,000 people dressed in historical costumes—including that of the samurai—parade through the city.

Craftsmen fashion the blade of a traditional samurai sword, *above.* The two swords that symbolized the samurai's status as a warrior were worn and used with much ritual.

that the samurai could not wear their two traditional swords in public—a tremendous blow to the samurai's sense of honor.

Although some samurai resigned themselves to their loss of status and found new occupations, many others were discontented and became involved in uprisings. In 1877, these samurai launched the Satsuma Rebellion, a large-scale revolt against the government, but suffered crushing defeat.

Nevertheless, remnants of the samurai tradition continued to endure until World War II, when more than 1,000 Japanese pilots flew *kamikaze* (divine wind) suicide missions against Allied warships. The kamikaze pilots, who crashed planes filled with explosives, considered it an honor to die for their emperor, believing that he ruled Japan by divine right. After the emperor surrendered in 1945, many Japanese committed hara-kiri in front of the Imperial Palace in Tokyo.

Kyoto

Emperor Kammu established Japan's capital at Kyoto on Honshu Island in A.D. 794. He called it *Heiankyo,* meaning *capital of peace and tranquility.* Many Japanese called it *Miyako,* meaning *imperial city,* or *Kyoto,* meaning *capital city.*

The first 400 years of the capital's history—from 794 to 1185—was a time of great cultural achievement called the Heian period. It marked the end of centuries of cultural borrowing from China and the beginning of distinct Japanese styles in literature, art, religion, and other areas, which differed greatly from those of the Chinese.

During the Heian period, the greatest literature was written by women. Sei Shonagon, a lady in waiting to a noblewoman, portrayed Heian court life around the early 1000's in her volume of essays called *The Pillow Book.* Another book about court life, *The Tale of Genji* by Murasaki Shikibu, is the finest achievement of Heian literature. Indeed, many consider this lengthy novel to be the greatest work of Japanese fiction. Written around 1010, Murasaki's book surpassed previous stories with its sophisticated style and insightful descriptions of human emotions.

Another important development of the early Heian period was the introduction of Japanese characters called *kana.* Until this time, the Japanese language had been written in Chinese characters. The use of kana allowed a new freedom of expression for the Japanese, and some of the Heian period's finest literary works were written in this new script.

Although Buddhism declined in China during this period, it flourished in Japan. Shinto, Japan's traditional religion, was merged to a great extent with Buddhism. Shintoists worship many gods, called *kami,* which they believe inhabit mountains, trees, and other elements of nature. During the Heian period, Shintoists identified Buddhist gods as kami and used Buddhist images to represent the kami in their

Japan has had three capitals since A.D. 710, *below,* when Nara was established as the seat of government. Kyoto replaced Nara in 794 and remained the capital until 1868. At that time, Emperor Mutsuhito moved the capital east to Tokyo. All three cities are on the island of Honshu.

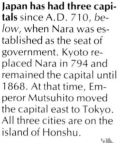

Modern Kyoto, *right,* retains some of the features of the ancient capital, including the layout of the city's streets.

1 Ryoan-ji Temple
2 Golden Pavilion
3 Old Imperial Palace (Gosho)
4 Sento Palace
5 Nijo-jo Castle
6 Pontocho
7 Yasaka Shrine
8 Nishi-Hongan-ji
9 Higashi-Hongan-ji
10 Shosei-en Garden
11 Kyoto National Museum
12 To-ji

Main street layout of ancient capital city known as Heiankyo when founded about A.D. 800

A Buddhist temple in Kyoto, *left,* set amid trees and gardens, combines the Buddhist principle of quiet contemplation with the Shinto reverence for nature.

Lovely gardens surround the teahouse of Kyoto's Imperial Palace, which was built in 794 and rebuilt in 1855.

The dance of the crane, a symbol of long life, is part of the Gion Festival held at Yasaka Shrine in Kyoto, *left.* The festival, which dates from the 800's, was originally held to honor a Shinto god of good health.

The Golden Pavilion, built by the shogun Yoshimitsu in 1397 and rebuilt in the 1950's, is one of the most beautiful sights in Kyoto. Gold leaf covers parts of the structure, which served as both a villa and a temple. The pavilion was reconstructed according to the original plan following a fire.

shrines. Kyoto became an important religious center, with many Buddhist temples and Shinto shrines housing priceless works of art.

In 1185, however, the Minamoto family gained control of the imperial court at Kyoto—an event that marked the end of imperial culture and ushered in nearly 700 years of military domination under shogun rule. Tokyo replaced Kyoto as the nation's capital in 1868.

Today, Kyoto is one of Japan's largest cities. Although Kyoto was the only major Japanese city to escape bombing during World War II, earthquakes, fires, and floods have destroyed most of the buildings, temples, and shrines built during the Heian period. However, many of these ancient structures have been rebuilt in their original style, and the layout of the city still preserves Kyoto's original grid pattern, with its intersecting north-south and east-west avenues. As a result, many modern streets follow the routes of the ancient city of Kyoto.

Tokyo

During most of its history, Tokyo was called *Edo*. Historians believe that a powerful family first lived in the area around 1180. Its location overlooking the wide Kanto Plain and Tokyo Bay gave the area military importance, and in 1457 a powerful warrior named Ota Dokan built a castle there. Edo developed around the castle.

However, Edo's development into Japan's chief city did not begin until 1590, when Ieyasu Tokugawa made it his headquarters. Edo became Japan's political center in 1603, after Ieyasu became shogun of Japan. When shogun control ended in 1868, Emperor Mutsuhito transferred the capital from Kyoto to Edo and moved into the Edo castle. Edo then was renamed *Tokyo*, which means *eastern capital*.

After 1868, Japan—and especially Tokyo—rapidly adopted Western styles and inventions. European architects and building styles were employed to develop the city, and little attempt was made to preserve wooden buildings constructed in the ancient Japanese architectural style. By the late 1800's, Tokyo had begun to look like a Western city.

The city had to be rebuilt following a massive earthquake in 1923, which destroyed most of central Tokyo. About 59,000 people died in the disaster. World War II brought destruction to Tokyo once again. More than 250,000 people were killed in the bombing or listed as missing, and about 97 square miles (251 square kilometers) of Tokyo were ruined. The people of Tokyo began to rebuild their city after the war, but without much planning. Buildings went up wherever there was room, and as a result many of Tokyo's structures are modern. Most of the remaining buildings in Japanese style are religious shrines or temples.

Tokyo today is Japan's major business and cultural center as well as the home of the Japanese emperor and the headquarters of the national government. The city is also a major tourist center. The Imperial Palace, the home of the Japanese emperor, and the Meiji Shrine, which lies about 3 miles (5 kilometers) southwest of the palace, draw tourists from all over the world. Many Japanese, dressed in traditional garments, visit the well-known shrine on New Year's Day. Tokyo's parks, with their spring displays of cherry blossoms and Japanese-style gardens, are also popular with tourists. They not only reflect the Japanese love of beauty, but also offer welcome relief from the crowds in a city that is now one of the most densely populated places on earth.

With about 8-1/3 million people, Tokyo is the fourth largest city in the world after Mexico City, Seoul, and Moscow. Tokyo, also called the *city proper,* is part of the *Metropolis of Tokyo*— a large metropolitan area that includes many communities west of the city and has a population of 11,618,281.

Tokyo's soaring population has created a serious housing shortage, and the city government has begun financing the construction of low-rent housing projects.

Shoppers throng the Nakamise-dori, a long, narrow avenue of shops, some of which date back to the 1700's and 1800's. This Tokyo street leads to the historic Kannon Temple in the Asakusa district. Reconstructed in the 1950's, the temple traces its origins back to the 600's.

Tokyo's Imperial Palace, *below right,* consists of several low buildings set in beautiful parklike grounds. Stone walls and a series of moats separate the palace from the rest of the city. The palace attracts thousands of visitors on January 2 and on the emperor's birthday, when it is open to the public.

Japan's *Shinkansen,* or bullet trains, make up a large fleet of high-speed, superexpress trains that run the length of the main island of Honshu.

The Metropolis of Tokyo, *maps below,* covers the same area as Tokyo Prefecture. (Prefectures are similar to states in the United States.) The city proper is the busiest, most heavily populated part of the Metropolis. The Imperial Palace lies at the heart of the city center. Northeast of the palace is the Kanda district, famous for its bookstores, and the Asakusa district, featuring many of the city's restaurants and theaters. Farther north lies popular Ueno Park, which offers such attractions as masses of cherry blossoms in spring, a concert hall, several museums and art galleries, a zoo, a temple and shrine built during the 1600's, and tombs of Japanese rulers. The Marunouchi district, Tokyo's business and financial center, lies east of the palace. The Ginza district, famous for its stores and nightclubs, is farther south.

However, most of Tokyo's housing developments are located far from the city proper, and workers who live in outlying areas spend up to four hours a day traveling to and from their jobs in downtown Tokyo. Severe traffic jams on the city's highways occur frequently, and every day nearly 10 million people cram aboard the city's commuter trains, which are so crowded that employees called *pushers* are hired to shove passengers into packed trains.

In spite of Tokyo's crowded conditions, the city has relatively little crime and poverty. Because of Tokyo's strong economy, most people can find jobs, and in addition the local and national governments provide aid for those who cannot support themselves.

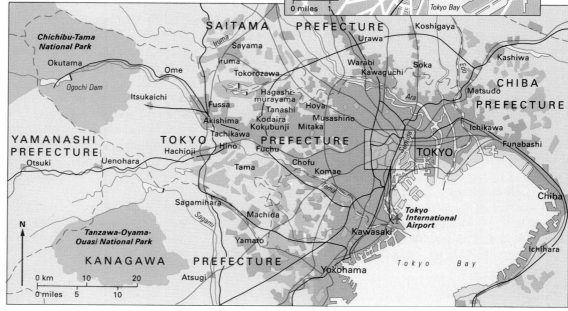

People

The Japanese people of today are probably descended from the Yayoi, a people who established agricultural villages in what is now Japan around the 200's B.C. Chinese, Koreans, and a group of people called the Ainu make up the largest minority groups in Japan. Some scientists believe that the Ainu were Japan's original inhabitants. Today, most of these people live on Hokkaido, the country's northernmost island.

More than 75 per cent of the Japanese people live in urban areas, and much of the population is concentrated in the four major metropolitan areas of Tokyo, Yokohama, Osaka, and Nagoya—all on the island of Honshu. Most Japanese city dwellers are employed in factories, businesses, the government, and service industries and enjoy a comfortable standard of living.

Only about 25 per cent of the Japanese people live in rural areas. Most of the rural population work as farmers, but some Japanese make their living by fishing and harvesting edible seaweeds along the coasts. Although their standard of living has increased steadily since World War II, most rural workers do not earn as much as city dwellers, and since the late 1950's many rural Japanese, particularly young people, have moved to urban areas to seek better-paying jobs.

East meets West

Life in Japan's big cities combines the old and the new, as well as the East and the West. High-rise office buildings dominate the commercial districts, and modern transportation systems carry millions of commuters each day. But many old customs still flourish. For example, numerous big-city shops specialize in traditional items, such as straw mats called *tatami,* which are used as floor coverings. In addition, even the most crowded cities have beautiful gardens, parks, and shrines—all of which reflect the traditional Japanese love of nature.

Although most of the country's city dwellers wear Western-style clothing in public, some Japanese people—particularly the older generation—still dress in the traditional kimono at home, and al-

Japanese businessmen in Western dress throng a Tokyo street. The Japanese economy is so strong that most people can easily find jobs. Some large corporations even guarantee their workers lifetime employment.

An elderly Ainu poses in traditional garb. While most of the Ainu have adopted the Japanese culture, some still follow their traditional way of life.

most all Japanese wear kimonos for festivals, holidays, and other special occasions. Worn by both men and women, the kimono is tied around the waist with a sash called an *obi.*

City housing, also a blend of East and West, includes both modern apartment buildings and traditional Japanese houses. Most traditional houses feature lovely gardens, graceful tile roofs, and sliding paper screens between rooms. Tatami cover the floors, and people sit on cushions and sleep on padded quilts called *futons.* However, many Japanese apartments and houses have one or more rooms with Western-style furniture and carpets on the floors.

Family life

Before 1945, many Japanese lived together in large extended families and followed strict social customs. Husbands had complete authority over their wives, and chil-

A businessman proudly displays pearls strung in his factory, *below*. The pearls come from oysters cultivated and harvested in Japan's coastal waters.

Children play in a snow-house, *below,* during the celebration of the Snow Festival in Sapporo, Hokkaido. Although Japanese parents are traditionally strict, a number of festivals are held especially for the children.

Seafood and rice constitute the traditional diet of the Japanese people. But meat and dairy products have been more popular since the 1950's, and many Japanese substitute bread for rice at some meals.

A New Year's parade goes on in spite of a snowstorm. Many Japanese dress for the occasion in their most elaborate and colorful kimonos.

Traditional garments are part of the wedding ceremony for many Japanese couples, *far left.*

dren were expected to obey their parents without question. Parents even selected their children's marriage partners—a bride and groom often met for the first time on their wedding day.

Today, most Japanese live in smaller family units consisting of only parents and children, and relationships within these families have become more democratic. Children are given much more freedom, and most young people select their own marriage partners. Women, too, guaranteed equal rights by the Constitution of 1947, are no longer dominated by their husbands or male relatives. As a result, an increasing number of Japanese women now work outside the home. Nevertheless, although the urbanization of Japan has produced many changes in the country, the Japanese people still maintain strong family ties and a deep respect for authority.

Ethnic and Social Groups

About 70,000 Chinese and about 675,000 Koreans make up the largest ethnic minority groups in Japan. Groups of Chinese and Koreans are scattered throughout Japan, though large Korean communities are established in Osaka, Kobe, and Kyoto. Koreans in Japan who have declared their loyalty to North Korea belong to an organization called Chosen Soren. This organization, which is supported by the North Korean government, runs its own schools where only the Korean language is spoken.

The Ainu people, totaling about 15,000, are another noteworthy minority group in Japan. Scientists are uncertain about the origin of the Ainu, who may have been Japan's original inhabitants. Some anthropologists think the Ainu are related to European peoples, while others believe they are related to Asian peoples or to the original inhabitants of Australia. Over the centuries, many Ainu have intermarried with the Japanese and have adopted Japanese customs. Today, only a small number of Ainu follow their traditional way of life. These Ainu live in isolated communities on Hokkaido Island.

Government aid

The Ainu have long been victims of discrimination, but they have started a movement to achieve fair treatment and the government has instituted a program to aid them economically. As a result of these measures, prejudice against the Ainu people has decreased.

The Korean minority in Japan also has faced discrimination. In 1990, however, government leaders from both South Korea and North Korea demanded equal treatment for the Koreans in Japan, and as a result Japan agreed to grant Koreans more civil rights.

Outcasts and criminals

The people who have suffered the greatest injustice in Japan are a group of Japanese known as the *burakumin* or *eta*. In Japanese feudal society, the burakumin were outcasts because they came from villages

The ramshackle houses of the burakumin are usually clustered in the poorer sections of Japan's urban areas. The buildings' inferior construction offers little resistance to the country's frequent earthquakes.

A vagrant stops for a rest outside a Tokyo shop, *top right*. Although most people in Japan can find jobs, those without the proper ancestral background may encounter discrimination and have difficulty finding work.

Shoe shining, *center right*, and other low-paying jobs, such as tanning leather and working in prisons and slaughterhouses, are often the only ones open to the burakumin.

Tattooed members of a yakuza clan parade down a Tokyo street, *bottom right*. The yakuza have a code of honor similar to that of the samurai, and violating the code may lead to physical punishment.

associated with tasks considered unclean by Buddhist doctrine. These tasks included the execution of criminals, the slaughter of cattle, and the tanning of leather. According to Buddhism, not only the tasks, but also the people who performed these duties, were unclean.

Feudalism was abolished in Japan in 1871, but the outcasts remain. Today, the burakumin number about 1 million, and though they are not ethnically different from other Japanese, they still suffer from discrimination. Many live in segregated urban slums or special villages. The bura-

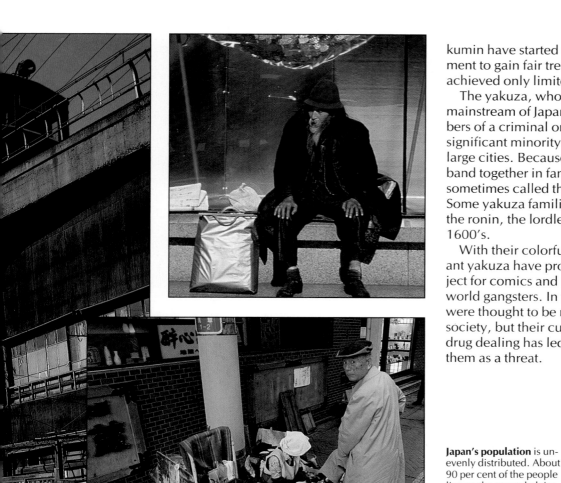

kumin have started an active social movement to gain fair treatment but have achieved only limited success.

The yakuza, who also live outside the mainstream of Japanese society, are members of a criminal organization and form a significant minority in Tokyo and other large cities. Because yakuza members band together in families or clans, they are sometimes called the "Japanese mafia." Some yakuza families claim kinship with the ronin, the lordless samurai of the 1600's.

With their colorful tattoos, the flamboyant yakuza have provided a popular subject for comics and movies about underworld gangsters. In the past, the yakuza were thought to be relatively harmless to society, but their current involvement with drug dealing has led the Japanese to regard them as a threat.

Japan's population is unevenly distributed. About 90 per cent of the people live on the coastal plains. The Pacific coast from Tokyo to Kobe is the most densely populated area. The Ainu have been pushed northward to Hokkaido, Japan's northernmost island.

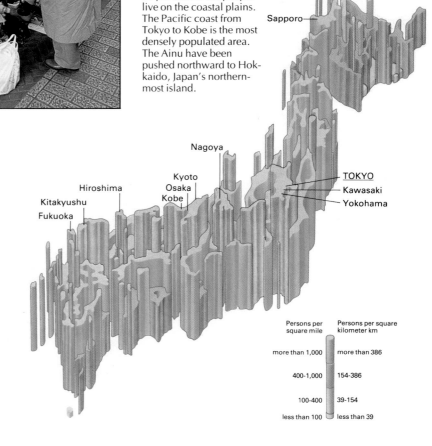

Sapporo

Nagoya

Hiroshima

Kyoto
Osaka
Kobe

Kitakyushu

Fukuoka

TOKYO
Kawasaki
Yokohama

Persons per square mile	Persons per square kilometer km
more than 1,000	more than 386
400-1,000	154-386
100-400	39-154
less than 100	less than 39

Popular Culture

Popular culture in Japan is a blend of Japanese traditions and modern Western influences. Since the Meiji period, Japan has borrowed sports and arts from the Western world, particularly the United States. However, much of this imported culture has been adapted to suit Japanese tastes. For example, baseball, which was introduced into Japan in the late 1800's, is now one of the country's most popular spectator sports. Yet to the Japanese, the game represents an art form in which play is carried out with order and harmony. The slowness of the game allows the spectators to pay attention to its every detail, and the duel between the pitcher and batter provides drama.

While baseball successfully combines Japanese spirit with Western know-how, sumo wrestling is uniquely Japanese. The country's national sport, sumo is immensely popular in Japan. The wrestlers perform on a raised platform, which accentuates their huge size, and many rituals surround the tournaments.

Among the participant sports in Japan, rubberball baseball, golf, skiing, and tennis are popular, as are *aikido, judo,* and *karate,* traditional Oriental martial arts that involve fighting without weapons. Another favorite pastime for Japanese of all ages is *pachinko,* a noisy type of pinball. A pachinko arcade contains hundreds of machines lined up in rows, and there are thousands of such arcades throughout Japan.

Motion pictures

Early Japanese films were heavily influenced by several forms of traditional Japanese drama. Silent films of the early 1900's employed a narrator who kept the audience informed of the plot—a custom derived from the Japanese *puppet* theater, which developed during the late 1600's. Some films borrowed the choruses and musical accompaniment of the *no* play, developed during the 1300's. Films featuring the stylized enactment of historical and domestic events were grounded in traditional *kabuki* theater, which began in the 1600's.

Samurai films, one of the most popular styles of film in Japan, evolved from the kabuki drama. Samurai films range from the artistic efforts of Akira Kurosawa, possibly the country's best-known film director, to action-packed swashbucklers that often combine swordsmanship with magic. However, the most popular films in Japan are movies about gangsters and professional gamblers—the so-called yakuza films. These films usually pit a "good" yakuza, who lives according to an established code of honor, against a gang of "bad" yakuza, who are willing to violate that code.

Baseball is a year-round obsession with the Japanese. Many enthusiasts have compared the concentration required to play the sport to the concentration of Zen Buddhists during meditation.

Perfecting their putting, Japanese businessmen exhibit their kinship with their Western counterparts. But unlike Western golfers, these players hone their skill on a practice area constructed on the roof of a Tokyo department store.

Popular music

Western styles have also influenced Japanese music. Popular songs in Japan, called *kayokyoku,* have borrowed heavily from many types of popular Western music, including pop, folk, rock, and heavy metal. However, the Japanese have developed their own brand of popular music sung by teen-age singers, usually female, called *kawaiko-chan,* meaning "cute" singers. The kawaiko-chan accompany their songs with hand gestures and body movements especially choreographed for them. These singers are discovered, packaged, and promoted by production companies.

A major influence on Japanese popular music was the invention of a technological innovation called *karaoke* (empty orchestra). In karaoke, a vocalist sings into a microphone while a prerecorded tape plays background instrumental music. Karaoke is used in bars, and video versions are popular in many Japanese homes.

A massive sumo wrestler, *below,* enters the ring while a sumo official, dressed in traditional costume, stands at attention. Sumo tournaments attract large, enthusiastic crowds, and many matches are televised.

Pachinko players, *below,* manipulate the machine's controls to direct a metal ball through a horizontal maze of obstacles. Winners are rewarded with extra balls that can be exchanged for prizes, or used to play again.

Imitating the latest Western fashions is popular among many young Japanese. Some even form groups that are identified by their tastes in clothing and music. Young men in one such group wear 1950's American clothing and pose and dance in public parks.

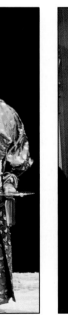

Shinto and Buddhism

Japan's oldest religion is Shinto—a native religion of Japan that dates from prehistoric times. *Shinto* means *the way of the gods,* and Shintoists worship many gods, or *kami,* which they believe are found in nature. The sun, water, mountains, trees, and stones have particular importance in Shinto worship. According to Shinto mythology, Japan's emperors are direct descendants of the sun goddess—the supreme kami. The other four elements of nature—water, mountains, trees, and stones—are prominent features throughout the Japanese landscape.

Shinto emphasizes rituals and ceremonies during which offerings are made to the kami, accompanied by prayers. On special occasions, festivals called *matsuri* are held to honor a particular god. Worshipers take part in solemn religious ceremonies during a matsuri, but they also entertain the god with food, drink, music, and dance, hoping to earn the god's good will. They believe that a god that is pleased by the festival will give them such gifts as long life, wealth, peace, good health, and a rich harvest.

These ceremonies and festivals generally take place at shrines, which range from a large complex of several buildings and gardens to a small space in someone's home. Roadside shrines dedicated to certain kami are also common. In addition, the kami themselves can be used as places of worship. Mountains, for example, have been worshiped as gods and used as religious shrines.

Shinto rituals and festivals celebrate the important occasions of life and of the agricultural year, such as birth, marriage, and harvesttime. Shinto expresses a simple joy in life, but it does not deal with death. Buddhism gave the Japanese new insight into this aspect of the life cycle.

Buddhism was introduced to Japan from China and Korea about A.D. 552, when Japan was heavily under the influence of Chinese ideas. Generally, Buddhism teaches that people can achieve peace and be free from suffering only if they rid themselves of attachment to worldly things. The Japanese accepted the religion, but added

some of their own appreciation of life and nature to the Buddhist teachings. They also merged Buddhist art and culture with Shinto beliefs, so temples and monasteries were built to honor Buddha, but Buddhist images were also used to represent Shinto gods.

A form of Buddhism called Zen, meaning *meditation,* was introduced into Japan in the 1100's. Zen Buddhists believe meditation is the key to achieving a state of spiritual enlightenment called *satori.* The development of the traditional Japanese tea ceremony was highly influenced by both

The ornate Toshogu Shrine at Nikko, built during the 1600's and dedicated to Shogun Ieyasu Tokugawa, contrasts sharply with the Japanese ideal of simplicity and harmony. During this period, the Japanese celebrated military power and material splendor.

Priests celebrate Aoi Matsuri, a major Shinto festival, with music, *right.* Participants in the festival, dressed like nobles of the late Heian period, lead a procession to two Shinto shrines, where ceremonies are held.

A wooden gate called a _torii_ is the symbol of Shinto, _left_. A torii consists of two posts connected by crossbars and stands at the entrance of a Shinto shrine. The posts represent pillars that support the sky, and the crossbars symbolize the earth. This torii, on a small island near Hiroshima, is partly submerged by the high tide.

The Great Buddha of Kamakura, _above_, serene and peaceful, sits cross-legged in meditation. Zen Buddhists also assume this position while meditating.

A Buddhist monk seeks donations on a busy street in Tokyo. Modernization has not resulted in the abandonment of religion in Japan. On the contrary, because Buddhism and Shinto reflect the Japanese ideal of a harmonious society working together, religion has helped to unify and strengthen the country.

Zen and Shinto. Zen turned the making and drinking of tea into a time of intense meditation, while Shintoists found beauty and significance in its simplicity.

During the 1800's, many Shintoists began to reject Buddhism. In the mid-1800's, a movement called _State Shinto_ stressed patriotism and the divine origins of the Japanese emperor. Japan's defeat in World War II shattered this movement. As a result, the government abolished State Shinto, and the emperor denied that he was divine.

Today, more than 60 per cent of Japan's population practices a combination of Shinto and Buddhism. For example, they may celebrate births and marriages with Shinto ceremonies, but funerals are observed with Buddhist ceremonies.

Japanese Gardens

The Japanese have been creating artfully landscaped gardens since before the A.D. 500's. These gardens range from spacious parks created for the nobility to tiny tea gardens. Large or small, the gardens mirror the natural beauty of Japan's landscape and reflect the Japanese ideals of simplicity, harmony, and tranquility.

A Japanese garden is designed to appear as if it were created by nature and not by a gardener's artistry, and Japanese gardens are often miniature replicas of scenes found in nature or described in Japanese literature. Thus, familiarity with a specific literary work, such as *The Tale of Genji*, might be required to fully appreciate the perfection of a particular Japanese garden.

Some Japanese gardens are quite elaborate, with stone paths winding past trees, buildings, hills, and lakes. These elements are cleverly arranged to present a new and pleasing scene at every turn in the path. Gardens in more confined spaces may give the illusion of spaciousness. The landscape gardener often uses perspective to achieve this end. For example, large and eye-catching objects may be placed in the foreground, while rocks and plants of decreasing size fill the background. The gardener may also conceal or camouflage the boundaries of the garden to disguise its actual size. Thus, as paths and streams disappear behind rocks or trees, they appear to be winding off into an undefined distance.

Japanese gardeners also use a technique known as "borrowed scenery," in which a rock or shrub is taken from its natural surroundings and placed in a garden setting. A large stone removed from a mountain, for example, may be set in a garden to create the feeling that the mountain is part of the scenery. Some gardens achieve a similar effect by making the surrounding scenery a part of the garden. In the Fukiage garden in Tokyo's Imperial Palace, for instance, the path winds past a terrace from which the visitor can view Mount Fuji.

Japanese gardens come in many forms, including the water garden, the dry garden, and the tea garden. The popularity of water gardens reflects the importance of water to the Japanese. In a water garden, lakes or ponds dotted with tiny islets sur-

round graceful buildings. Nestled in a forest of trees, Kyoto's Golden Pavilion actually sits on a lake. The reflection of the pavilion in the water seems to draw nature into the building and draw the building into nature.

Among the many different kinds of Japanese gardens, the dry garden is perhaps the most unusual to Western eyes. Unlike the Western garden, with its colorful flower beds and grassy lawns, a dry garden contains only sand and rocks. The rocks, carefully chosen for their shape, color, and size, are arranged on sand that has been raked in geometric patterns.

The stone garden at Ryoanji Temple in Kyoto contains only rocks and sand, *far left*. The sand is raked to represent the waves of the sea, and the rocks symbolize islands emerging from the water.

Trees surrounding a lake on the grounds of the Imperial Palace in Kyoto are the result of nature and the gardener's artistry. The trees have been carefully trimmed to create the most pleasing composition of shapes, color, and foliage.

The gardens in Kyoto's Shugakuin Palace, *left,* reflect the harmonious relationship between gardens and architectural structures. Screen doors on every side of the building open onto the garden, so that the garden appears to enter the building.

Celebrating the appearance of the cherry blossoms and the coming of spring, Japanese women, dressed in traditional costumes with sashes the color of cherry blossoms, parade through a Tokyo street, *below left*. Flower-viewing parties are held throughout Japan during this period, and many people picnic beneath the lovely cherry blossoms.

Many dry gardens re-create a landscape in miniature. For example, the swirling patterns of sand in the dry garden at Daisenin Temple in Kyoto are designed to suggest a waterfall and an ocean, while the rocks set against the sand symbolize boats and bridges. By contrast, the famous rock garden at Ryoanji Temple in Kyoto is abstract in design, with 15 rocks set like islands on waves of white, raked sand. The dry garden creates a simple world of stillness and beauty that reflects the Buddhist ideal of serenity. It also draws on the Shinto belief in the spiritual forces of nature.

The tea garden, also inspired by Buddhism, is built at the approach to a Japanese teahouse, where a ceremonial cup of tea inspires quiet repose and refreshes the spirit. A traditional tea garden is very simple in order to avoid distracting the mind from the tea ceremony. Many of these gardens consist only of a short path of stepping stones leading to the teahouse. The stones, chosen for their size, shape, and texture, are placed close together along the path so that a person must concentrate on walking across the stones and is, therefore, forced to disregard the surroundings and concentrate on the ceremony.

Jordan

Jordan is an Arab kingdom in the heart of the Middle East. Much of Jordan's modern history has been shaped by events in the land to its west, the area once called Palestine. In fact, during the early 1900's, Jordan was called *Transjordan* because of its location across the River Jordan from Palestine.

But the history of the region that is now Jordan goes back thousands of years. About 2000 B.C., Semitic nomads moved into the region, and by about 1200 B.C., four Semitic peoples—Ammonites, Amorites, Edomites, and Moabites—made a living there as farmers and traders. During the 900's B.C., the Israelite kings David and Solomon conquered the region, but Moabites led by King Mesha regained control about 50 years later.

A series of foreign invaders and rulers followed. In the 400's B.C., the area was controlled by the Nabataeans—traders who carved a capital city out of the rose-colored cliffs at Petra. In the 60's B.C., the Romans conquered the region. When the Roman Empire split in the late A.D. 300's, Jordan became part of the Byzantine Empire.

Muslims from the Arabian Peninsula defeated Byzantine armies in 636. The conquering Arabs brought the religion of Islam and the Arabic language to the people of the area. The Arabs established an important pilgrimage route through Jordan to Mecca, their holy city in Arabia.

About 1100, Christian crusaders from Europe captured land in the Middle East, including parts of Jordan, but the Muslim leader Saladin drove out the crusaders in 1187. Today, Saladin's shield, helmet, and eagle are displayed on Jordan's coat of arms.

Jordan was part of the Ottoman Empire during World War I (1914–1918) when Sherif Hussein of Saudi Arabia led an Arab revolt against Ottoman rule. With the help of Great Britain, the Arabs defeated the Ottomans in the Middle East.

The United Kingdom was then given the right to administer lands east and west of the River Jordan. In 1921, the land east of the river was named Transjordan and given partial self-government, with Hussein's son, Abdullah, ruling as *emir* (prince). The nation gained complete independence in 1946, when it was renamed Jordan.

Jordan became involved in the affairs of Palestine—the land west of the River Jordan—

in 1948. In that year, part of Palestine became the new state of Israel, which was created as a homeland for Jewish people. But as soon as the new state was proclaimed, Jordan and other Arab countries attacked Israel.

When the fighting ended in 1949, Israel occupied much of Palestine. Jordan took over the West Bank of the River Jordan and east Jerusalem. But it gained people as well as land. Jordan's population more than tripled with the addition of about 400,000 Palestinian Arabs who lived on the West Bank and about 450,000 Palestinian refugees who moved from Israel into Jordan. This large Palestinian population caused political and economic problems. The Palestinians required food and shelter, and they competed with the Jordanians for power.

In 1951, Palestinians assassinated Abdullah. In 1953, Abdullah's grandson, Hussein, took control as Jordan's king.

Increasing Arab-Israeli tensions led to the formation of the Palestine Liberation Organization (PLO) in 1964. Israel and the PLO staged raids against each other from time to time. Finally, on June 5, 1967, Israel attacked Jordan's ally Egypt. Jordan responded by attacking Israel. During the Six-Day War that followed, Jordan lost control of east Jerusalem and the West Bank. In addition to the influx of about 300,000 Palestinians from the West Bank, Jordan suffered from the loss of farmland and tourist income.

After the 1967 war, many Palestinians formed guerrilla groups to fight Israel. By early 1970, these forces had become almost a second government in Jordan and threatened to overthrow the king.

On Sept. 17, 1970, the Jordanian Army attacked the Palestinian guerrillas and defeated them within a month. But isolated fighting continued, and Jordan was drained by the battles. At a meeting of Arab leaders in 1974, King Hussein agreed that the West Bank should become an independent Palestinian state if Israel withdrew from that region. Nevertheless, Jordan continued to help fund public services in the West Bank until 1988. In July 1994, Jordan and Israel signed a declaration that formally ended war between the two countries.

Hussein died in 1999 and was succeeded by his son Abdullah.

Jordan Today

Jordan is a land of many contrasts, where modern cities rise near ancient ruins. And, like its landscape, its government is a mix of past and present. Since 1946, the country has been an independent constitutional monarchy.

Government

Jordan is ruled by a king with wide powers. The king appoints a prime minister to head the government. He also appoints the members of the Council of Ministers, Jordan's cabinet, and the members of the Senate— one house of the National Assembly, Jordan's legislature. Senators serve four-year terms.

The second house of the legislature is the House of Representatives. Deputies are elected for four years, but the recent former king, Hussein, twice dismissed the entire National Assembly. In addition, the king names a governor to head each district in Jordan and appoints all the judges in the country.

The government owns and operates the nation's radio and TV stations and closely controls all communications. Nevertheless, there is more freedom of speech in Jordan than in many other Middle East countries.

Economy

Jordan has a developing economy based on free enterprise. The nation depends on for-eign aid and on the economy of the Middle East region as a whole. Many Jordanians work outside the country in neighboring oil-producing Arab nations and send money home to their families in Jordan.

Within Jordan, service industries make up the largest part of the economy, employing about 70 per cent of the country's workers and accounting for more than 65 per cent of the total value of the nation's economic production. Many service workers are employed by the government, while others work in banking, education, insurance, trade, tourism, or transportation.

Many Jordanian people work for the military. In the face of wars with Israel in the 1960's and 1970's, as well as threats from Palestinians within the country, Jordan's government maintains large armed forces. But the cost of doing so is high and places a burden on Jordan's economy.

Jordan has few natural resources. Workers mine phosphates and potash, but the country lacks the valuable oil found in neighboring Arab lands. Jordan has only one dam for producing its own electricity and so must import oil to meet its energy needs. And because Jordan is mostly desert, only 3 per cent of the land is farmed.

Together, manufacturing and mining employ only about 12 per cent of Jordanian

---

FACT BOX

COUNTRY

Official name: Al Mamlakah al Urduniyah al Hashimiyah (Hashemite Kingdom of Jordan)
Capital: Amman
Terrain: Mostly desert plateau in east, highland area in west; Great Rift Valley separates East and West Banks of the Jordan River
Area: 34,445 sq. mi. (89,213 km²)

Climate: Mostly arid desert; rainy season in west (November to April)
Main river: Jordan
Highest elevation: Jabal Ramm, 5,755 ft. (1,754 m)
Lowest elevation: Dead Sea, 1,310 ft. (399 m) below sea level

GOVERNMENT

Form of government: Constitutional monarchy
Head of state: Monarch
Head of government: Prime minister
Administrative areas: 12 muhafazat (governorates)

Legislature: Majlis al-'Umma (National Assembly) consisting of the Senate with 40 members serving four-year terms and the House of Representatives with 80 members serving four-year terms
Court system: Court of Cassation, Supreme Court (court of final appeal)
Armed forces: 104,000 troops

PEOPLE

Estimated 2002 population: 6,639,000
Population growth: 3.1%
Population density: 176 persons per sq. mi. (68 per km²)
Population distribution: 78% urban, 22% rural
Life expectancy in years: Male: 75 Female: 80
Doctors per 1,000 people: 1.7
Percentage of age-appropriate population enrolled in the following educational levels: Primary: 69 Secondary: 66 Further: 9

0 km 100 200
0 miles 100

SYRIA

IRAQ

Syrian Desert

Yarmuk
Irbid   Ar Ramtha
Janin   Al Mafraq
Nablus
Zarqa   Jabal el Ashaqif
WEST BANK
Ram Allah   As Salt   Az Zarqa
Jericho   AMMAN
Jerusalem   Madaba   Azraq ash-Shishan
Bethlehem   Dead Sea   Dabah
Hebron   -1,310 ft (-399 m)
Al Maziaah
ISRAEL   Al Qatranah
Al Karak
Wadi al Hasa   Ard as Sawwan
At Tafila   Al Hasa   Bair
Ash Shawbak   Jabal al Adhiriyat
Al Jafr
Ash Sharah   Maan
Ras an Naqb.

SAUDI ARABIA

Al Aqabah   Jabal Ramm
5,755ft (1,754m)   Ar Ramlah
EGYPT   Al Mudawwarah

N

Gulf of Aqaba

Jordan, an Arab kingdom in the Middle East, borders Syria, Iraq, Saudi Arabia, Israel, and the West Bank—a territory once controlled by Jordan but lost to Israel in 1967. Amman is Jordan's capital and largest city.

Jordanian police, *right,* patrol the desert town of Wadi Rum on camelback. Like many Middle East countries, much of Jordan's land is desert.

A Palestinian refugee living in Jordan gazes out his window, perhaps pondering his future.

Languages spoken:
Arabic (official)
English
Religions:
Sunni Muslim 96%
Christian 4%

TECHNOLOGY

Radios per 1,000 people:
372
Televisions per 1,000 people: 84
Computers per 1,000 people: 22.5

ECONOMY

Currency: Jordanian dinar
Gross national income (GNI) in 2000:
$8.4 billion U.S.
Real annual growth rate (1999–2000):
3.9%
GNI per capita (2000): $1,710 U.S.
Balance of payments (2000):
$59 million U.S.
Goods exported: Phosphates, fertilizers, potash, agricultural products, manufactures
Goods imported: Crude oil, machinery, transport equipment, food, live animals, manufactured goods
Trading partners: Iraq, Germany, India, Saudi Arabia, United States

workers and account for about 20 per cent of the nation's total economic production. Large plants include a petroleum refinery and fertilizer and cement factories. Smaller plants produce batteries, ceramics, cigarettes, detergents, food products, pharmaceutical products, shoes, and textiles.

With the help of modern farming methods, many farmers along the River Jordan grow citrus fruits and vegetables. East of the river valley, farmers cultivate such grains as barley and wheat and such fruits as grapes and olives as well as vegetables and nuts.

Jordan has a good transportation system with paved highways that link the country to all its neighbors. The country is landlocked except for a tiny stretch of land in the southwest, on the Gulf of Aqaba. There, Al Aqabah, Jordan's only port, has been well developed to handle cargo.

Land and People

Jordan has a varied landscape of deserts, mountains, deep valleys, and rolling plains. The nation's land area is divided into three main regions: the Jordan River Valley, the Transjordan Plateau, and the Syrian Desert.

The Jordan River Valley, a deep, narrow valley near the western border, extends along the River Jordan from the Sea of Galilee to the Dead Sea. The valley is actually part of the Great Rift Valley, a deep depression in the earth's surface that extends well into Africa.

Because the Jordan River Valley receives very little rain, it was generally unsuitable for farming until the 1960's. At that time, the country developed an irrigation system and began to use plastic-covered hothouses to grow fruits and vegetables. These modern methods have made the river valley Jordan's major agricultural area.

East of the Jordan River Valley and the Dead Sea, the land rises steeply to form the Transjordan Plateau. This wedge-shaped plateau begins at Jordan's northern border, narrowing as it extends southward to the region around Maan. The high, rolling plains of the Transjordan Plateau are cut by steep *wadis,* or dry valleys. This area includes Jordan's largest cities as well as most of its farmland.

East and south of the plateau lies the Syrian Desert, Jordan's third major region. Hot and dry, the Syrian Desert is the northern part of the vast desert area that covers much of the Arabian Peninsula.

Most of Jordan's approximately 5 million people live in the northwest on the fertile high plateau. About 60 per cent of the people are native Jordanian Arabs, with a few small ethnic groups of Armenian Christians and Circassian Muslims. The remaining 40 per cent of the people are mainly Palestinian Arabs.

Most of these Palestinians fled to Jordan as refugees after the Arab-Israeli wars of 1948 and 1967. Others moved from the West Bank to the capital city of Amman between these wars. About 10 per cent of Jordan's people live in crowded refugee camps set up by the United Nations (UN), which also operates schools for the refugees.

The great majority of Jordanians—about 95 per cent—are Muslims; almost all follow the Sunni, or orthodox, branch. Islam affects every aspect of life for Jordanian Muslims.

Devout Muslims pray five times a day, attend a mosque, fast, give to the poor, and make a pilgrimage to the sacred city of Mecca in Saudi Arabia.

While Arabic is the official language of Jordan, English is widely taught and spoken, and the government prints many documents in both languages. Ethnic minorities often speak their own language.

About 78 per cent of Jordan's people live in towns or cities. Almost all the houses and apartments have electricity and running water, but some Jordanians have to live in dense, crowded neighborhoods. However, living conditions in Jordan are generally better than in many other developing countries.

Most rural Jordanians live in village houses made of stone and mud or concrete. Many villagers grow crops and raise goats and chickens, while others work in construction or mining.

A Bedouin entertains his visitors with music and tea at an encampment in the desert. Since the 1950's, many of these nomadic people have settled in rural villages or towns. Today, Jordan is largely an urban society.

The ancient city of Gerasa at present-day Jarash, near Amman, became an important Roman trading center after the A.D. 60's. Towering columns and stone avenues remind visitors of Jordan's 300 years as a Roman colony.

Less than 5 per cent of Jordan's people are Bedouins—Arab nomads who live in tents and roam the desert with their camels and sheep in search of water and pasture. But since the mid-1900's, many of these nomadic people have settled in towns and villages.

Most Jordanians wear modern Western clothing, but men often cover their heads with a cloth called a *kaffiyeh,* and many women wear long, loose-fitting dresses. Some rural Jordanians, including the Bedouins, wear traditional robes.

Jordanians are very social people who enjoy getting together at large family gatherings and picnics. They eat a variety of foods, including cheese, cracked wheat, flat bread, rice, vegetables, and yogurt. Chicken is popular, as is lamb, which is cooked in yogurt and served on a large tray of rice in a traditional Jordanian dish called *mansef.*

More than 1 million Palestinians live in Jordan, and many others have taken refuge in surrounding countries.

The rolling plains of the Transjordan Plateau, *above,* provide most of the country's farmland, though the Jordan River Valley is now a more productive agricultural area.

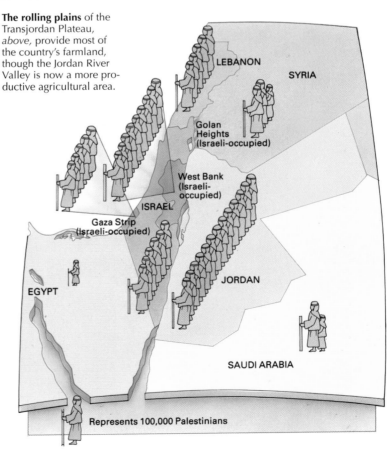

LEBANON
SYRIA
Golan Heights (Israeli-occupied)
West Bank (Israeli-occupied)
ISRAEL
Gaza Strip (Israeli-occupied)
JORDAN
EGYPT
SAUDI ARABIA
Represents 100,000 Palestinians

771

Kazakhstan

Kazakhstan is an independent country bordered by China in the east; Turkmenistan, Uzbekistan, and Kyrgyzstan in the south; Russia and the Caspian Sea in the west; and Siberia (part of Russia) in the north.

History

Turkish tribes originally settled the region that is now Kazakhstan. In the 1200's, they were conquered by the Mongols, who established *khanates*—realms ruled by Mongol chieftains—that dominated the region until the Russian conquest began in 1730.

The Russians set up forts throughout the land and established the capital of Verny, now called Almaty. In 1920, the Kazakh Republic was formed as an *autonomous* (self-governing) republic, and as an autonomous republic of Russia, it joined the Soviet Union in 1922. In 1936, Kazakhstan became a separate union republic known as the Kazakh Soviet Socialist Republic.

In 1990, Kazakhstan declared that its laws took precedence over those of the Soviet Union. Unlike most of the other Soviet republics, Kazakhstan did not immediately declare its independence following the upheaval in the Soviet Union sparked by the failed coup against Soviet President Mikhail Gorbachev's government in August 1991.

In December 1991, Belarus, Russia, and Ukraine announced they were establishing a loose confederation of former Soviet republics called the Commonwealth of Independent States (CIS). Kazakhstan declared its independence the following week. Also during that week, Gorbachev and Russian President Boris Yeltsin agreed to dissolve the Soviet Union by the end of the year and replace it with the CIS. Kazakhstan agreed to join the CIS, becoming the second largest country in the new commonwealth.

On Dec. 21, 1991, the presidents of 11 former Soviet republics met in the Kazakh capital of Almaty to sign a declaration wherein the member countries of the CIS agreed to respect one another's borders and independence and harness their nuclear force under a unified command. The agreement became known as the Almaty Declaration.

In 1997, Kazakhstan moved its capital from Almaty to the city of Akmola, now called Astana.

Land and people

Before the Bolshevik Revolution of 1917, the Kazakhs were a nomadic people who roamed the land grazing their livestock. Because they had no written language, the

FACT BOX

COUNTRY

Official name: Qazaqstan Respublikasy (Republic of Kazakhstan)
Capital: Astana
Terrain: Extends from the Volga to the Altai Mountains and from the plains in western Siberia to oases and desert in Central Asia

Area: 1,049,155 sq. mi. (2,717,300 km²)
Climate: Continental; cold winters and hot summers, arid and semiarid
Main river(s): Syr Darya, Irtysh, Ural
Highest elevation: Mt. Tengri, 20,991 ft. (6,398 m)
Lowest elevation: Karagiye Depression, 433 ft. (132 m) below sea level

GOVERNMENT

Form of government: Republic
Head of state: President
Head of government: Prime minister
Administrative areas: 14 oblystar, 3 qala (cities)

Legislature: Parliament consisting of the Senate with 47 members serving six-year terms and the Majilis with 77 members serving five-year terms
Court system: Supreme Court, Constitutional Council
Armed forces: 65,800 troops

PEOPLE

Estimated 2002 population: 16,191,000
Population growth: -0.05%
Population density: 15 persons per sq. mi. (6 per km²)
Population distribution: 56% urban, 44% rural
Life expectancy in years: Male: 58 Female: 69
Doctors per 1,000 people: 3.5
Percentage of age-appropriate population enrolled in the following educational levels: Primary: 97 Secondary: 87 Further: 23

KAZAKHSTAN

Kazakhs had no books, newspapers, or schools.

The Soviet "Virgin Lands" project of the 1950's and 1960's changed Kazakhstan forever. This agricultural project transformed the vast steppes of Kazakhstan into cropland. As a result, thousands of people from all over the Soviet Union—particularly Russia—migrated to Kazakhstan to work in the fields.

Although the Virgin Lands project was largely a failure, many immigrants settled in Kazakhstan permanently. As a result, most of the people living in Kazakhstan today belong to the Russian nationality group.

Kazakhstan now has an economy based on industry and agriculture. Cotton, barley, and wheat are grown in the south, and farmers raise sheep and cattle throughout the country. Rich deposits of chromium, coal, copper, iron ore, lead, nickel, petroleum, and zinc have contributed to the republic's industrial development.

Because Kazakhstan is thinly populated, the Soviet government chose to locate their first nuclear-test site at Semipalatinsk in northeastern Kazakhstan. Soviet scientists launched their first artificial Earth satellite from Baikonur, in central Kazakhstan.

The Tian Shan Mountains provide a backdrop for the former capital city of Almaty, which means *Father of Apples*—so named because it was built on the site of a village known for its apples.

Languages spoken:
Kazakh (Qazaq, state language) 40%
Russian (official) 66%

Religions:
Muslim 47%
Russian Orthodox 44%
Protestant 2%

TECHNOLOGY

Radios per 1,000 people: 422

Televisions per 1,000 people: 241

Computers per 1,000 people: N/A

ECONOMY

Currency: Kazakhstani tenge

Gross national income (GNI) in 1998: $18.8 billion U.S.

Real annual growth rate (1999–2000): 9.6%

GNI per capita (2000): $1,260 U.S.

Balance of payments (2000): $1,074 million U.S.

Goods exported: Oil, ferrous and nonferrous metals, machinery, chemicals, grain, wool, meat, coal

Goods imported: Machinery and parts, industrial materials, oil and gas, vehicles

Trading partners: Russia, European Union, China, Ukraine, United States

Kazakhstan is a vast wilderness of broad, sandy deserts and low, grassy plateaus.

Kenya

The country of Kenya on the east coast of Africa attracts thousands of tourists each year. Kenya's beautiful landscape and spectacular variety of wildlife draw visitors from around the world.

Land and wildlife

Kenya's coast is lined with beautiful sandy beaches and lagoons. Mangrove swamps, cashew trees, coconut palms, and patches of tropical rain forest thrive in the hot, humid climate.

Inland, Kenya consists of dry, grassy plains with no cities or towns. Nomads roam this region in search of pasture and water for their animals.

In southwestern Kenya, the highland region has lofty mountains, fertile valleys, and grassy plateaus. There, Mount Kenya rises more than 17,000 feet (5,000 meters) high, and the Great Rift Valley, formed by cracks in the earth's surface, cuts across the highland.

Kenya's highland and plains regions are home to a fascinating variety of animals. Antelope, buffalo, cheetahs, elephants, giraffes, leopards, lions, rhinoceroses, and zebras roam the open country. Large ostriches and elegant storks add grace to the landscape, and crocodiles and hippopotamuses are found where water is plentiful.

History

Kenya has a rich and varied human history as well. Scientists believe that the remains of human beings found in Kenya may be 2 million years old. About 3,000 years ago, people from other parts of Africa began moving into what is now Kenya. Arabs began visiting the coast about 2,000 years ago and eventually established settlements there.

In the early 1500's, the Portuguese came to the Kenyan coast and took control of the area from the Arabs. In the 1600's, the Arabs defeated the Portuguese and regained control. However, these events had little impact on the people living in the interior plains or the highland area.

Then in the 1800's, the British became interested in the region, and in 1887 British businessmen leased part of the coast from the sultan of Zanzibar. The British government took over the coastal area in

1895 and soon extended its control to all of Kenya, which became known as British East Africa.

Britain completed a railroad between the port city of Mombasa and Lake Victoria in 1901 and encouraged Britons and other Europeans to settle in the area. Soon, Europeans owned large Kenyan farms and hired Africans to work for them. The Africans had no say in their government.

During the 1940's, the Kikuyu people of central Kenya and other Africans began opposing British rule—many Kikuyu were living in poverty under the British. The Kikuyu and other Kenyans formed a political party called the Kenya African Union (KAU) in 1944, and a Kikuyu named Jomo Kenyatta became the party's leader in 1947.

During this time, a secret movement—sometimes called *Mau Mau*—developed among Kikuyu KAU members. The Mau Mau sought greater unity among Kenyan Africans and demanded new British policies and programs designed to improve Africans' lives.

In 1952, when Mau Mau members began committing terrorist acts to further their cause, the British jailed thousands of them. Fighting then broke out. Kenyatta, convicted of leading the Mau Mau, was jailed in 1953, but the fighting continued for three more years. By 1956, about 13,500 Africans, 95 Europeans, and 29 Asians had been killed.

During the late 1950's, all of Kenya's Africans began demanding self-rule. Britain agreed to the demand, and in February 1961, elections were held for a new parliament. Kenyatta's political party, called the Kenya African National Union (KANU), won the elections, but the party refused to take office until their leader was released from jail. However, Britain refused to release Kenyatta until August of that year. In the meantime, a rival party, the Kenya African Democratic Union (KADU), formed a government.

Full independence was finally gained on Dec. 12, 1963, and Kenyatta became prime minister. When Kenya was made a republic in 1964, Kenyatta's title was changed to president, and he remained in that office until his death in 1978.

Kenya Today

Although Kenya has made much economic progress since it became independent, the country faces major problems. Only about 20 per cent of the land is suitable for farming, but the population is growing at a rapid rate. Today, finding ways to feed its growing population is Kenya's chief challenge.

Since independence, Kenya has greatly increased its industry and tourist trade. Some of the money and machinery needed for the new industries has come from foreign investors.

Some Kenyans object to this foreign investment because they are afraid it gives too much influence over their country to outsiders. They also object to tourism because that industry also relies on money from outsiders. These Kenyans compare their country's current situation to its colonial past, when Kenya's economic and cultural life was dominated by non-Africans. However, other Kenyans support the growth in industry and tourism as a way to improve the nation's economy—and therefore, the people's lives.

Large numbers of Kenyan parents value education as the key to a better life for their children. The vast majority send their children to school—at least to elementary

The graceful Jamaa Mosque, *above,* is one of many important buildings in the central district of Nairobi, Kenya's capital and largest city. Although Nairobi is also a major commercial center, the city has a national park within its borders where lions and other wild animals roam the open land.

school—even though the law does not require them to do so.

Because of parents' demands, Kenya's government has greatly increased the number of schools and now operates schools in most sections of the country. Also, in many places

FACT BOX

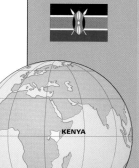

KENYA

COUNTRY

Official name: Republic of Kenya
Capital: Nairobi
Terrain: Low plains rise to central highlands bisected by Great Rift Valley; fertile plateau in west
Area: 224,962 sq. mi. (582,650 km²)
Climate: Varies from tropical along coast to arid in interior

Main rivers: Athi/Galana, Tana
Highest elevation: Mount Kenya, 17,058 ft. (5,199 m)
Lowest elevation: Indian Ocean, sea level

GOVERNMENT

Form of government: Republic
Head of state: President
Head of government: President
Administrative areas: 7 provinces, 1 area

Legislature: Bunge (National Assembly) with 222 members serving five-year terms
Court system: Court of Appeal, High Court
Armed forces: 24,200 troops

PEOPLE

Estimated 2002 population: 31,069,000
Population growth: 1.53%
Population density: 139 persons per sq. mi. (54 per km²)
Population distribution: 80% rural, 20% urban
Life expectancy in years: Male: 47 Female: 49
Doctors per 1,000 people: 0.1
Percentage of age-appropriate population enrolled in the following educational levels: Primary: 92 Secondary: 31 Further: 1

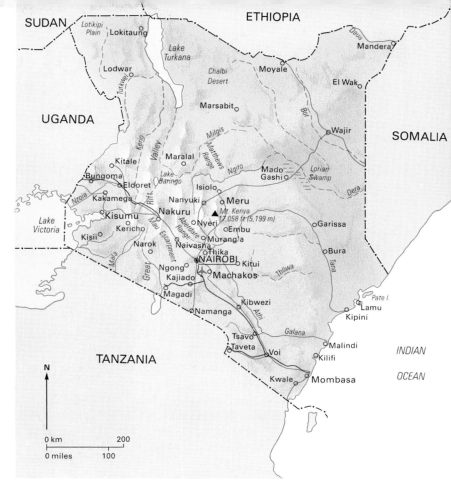

Kenya, a country on the east coast of Africa, extends deep into the interior of the continent. The equator runs through the center of Kenya.

without government schools, the people have set up their own self-help, or *harambee*, schools. *Harambee* is a Swahili word that means *pulling together*.

Kenya today is a republic headed by a president. Its Constitution, adopted in 1963, grants the people such rights as freedom of speech and religion. Kenyan citizens 18 years old or older may vote in elections.

Languages spoken:
English (official)
Kiswahili (official)
Numerous indigenous
languages
Religions: Protestant 38%
Roman Catholic 28%
indigenous beliefs 26%
Muslim 7%

TECHNOLOGY

Radios per 1,000 people:
223

Televisions per 1,000 people: 25

Computers per 1,000 people: 4.9

ECONOMY

Currency: Kenyan shilling
Gross national income (GNI) in 2000: $10.6 billion U.S.
Real annual growth rate (1999–2000): -0.2%
GNI per capita (2000): $350 U.S.
Balance of payments (2000): -$238 million U.S.
Goods exported: Tea, coffee, horticultural products, petroleum products
Goods imported: Machinery and transportation equipment, petroleum products, iron and steel
Trading partners: United Kingdom, Uganda, Tanzania, United Arab Emirates, United States

The voters elect the president and 210 members of the National Assembly—Kenya's legislature—all to five-year terms. Candidates running for president must run for a seat in the National Assembly at the same time and must win both elections to become president.

The president appoints 12 additional members to the National Assembly and also selects about 20 Assembly members to serve as Cabinet ministers. Each Cabinet minister heads an executive department of the government, and one Cabinet member also becomes vice president.

In 1982, Kenya's leaders changed its Constitution to outlaw all political parties except the Kenya African National Union. In 1991, the Constitution was amended to allow for a multiparty system. In Kenya's first multiparty elections, held in late 1992, Daniel arap Moi scored a narrow victory over eight challengers.

People and Economy

Kenya's population is more than 30 million and until recently was growing at one of the world's fastest rates. Fortunately, Kenya's population growth has slowed to an average rate. Small numbers of Asian Indians, Europeans, and Arabs live in Kenya, but about 99 per cent of Kenya's people are black Africans.

The black Africans belong to about 40 different ethnic groups. The largest group, the Kikuyu, make up about 20 per cent of the entire population, while four other ethnic groups—the Kalenjin, Kamba, Luhya, and Luo—each make up from 10 to 15 per cent of the population. The Maasai, a well-known ethnic group, are nomads who tend herds of cattle on the vast plains of Kenya.

Kenya's ethnic groups speak different languages and may even follow different ways of life. These differences—as well as differences in economic and social development—have sometimes led to tension between the groups, but the Kenyan government has tried to overcome these ethnic divisions and create a sense of national unity.

The Kiswahili language, which is spoken by large numbers of Kenyans in addition to their local language, helps unite the people. Kiswahili is Kenya's national language and is widely used between people of different ethnic groups. English, the official language, is spoken by most educated Kenyans.

Kenya has a developing economy that relies heavily on agriculture, but manufacturing is becoming more important. Together, manufacturing and construction account for about a fifth of Kenya's economic production. Service industries such as government, wholesale and retail trade, and tourism account for most of the rest.

Rural life

About 75 per cent of the people live in rural areas, mainly in farm settlements made up of small houses with thatched roofs, mud or stick walls, and dirt floors. They raise crops and livestock for a living.

Corn, the basic food crop, is often ground into meal to make porridge and used to make stew. Other food crops include bananas, beans, cassava, potatoes, sweet potatoes, and wheat. Many farmers struggle to produce enough food for themselves, but some have extra crops or beef and milk to sell.

Workers on a tea plantation in the highland region harvest one of Kenya's main cash crops. Although most Kenyan farms are small, coffee and tea also are grown on huge estates.

A herdsman tends his cattle, *right,* in southwestern Kenya. Most of Kenya's rural people raise crops and livestock for a living. Some must also work at other jobs to support their families.

Many Kenyan farmers work at part-time jobs for added income. Some work as blacksmiths, carpenters, shoemakers, or tailors. Others work part-time on large farm estates—especially coffee and tea plantations—that are owned by wealthy landowners.

Coffee and tea are *cash crops*—that is, crops raised for export. Coffee is Kenya's chief cash crop and the nation's most important source of income. Other cash crops include cashews, cotton, pineapples, sugar cane, *pyrethrum* (used to make insecticide), and *sisal* (fiber used to make rope). These crops account for about a third of Kenya's economic production.

Maasai people perform a traditional dance. These tall, slender nomads are famous for their skill in using weapons and their strong, independent ways. The Maasai move from place to place in search of grazing land and water for their animals. They rely on their animals for milk and other food and judge people's wealth by the number of animals they own.

A nomads' camp, made up of a few simple huts, shelters livestock herders in the desertlike region near Lake Turkana, at Kenya's northern border.

Urban life and tourism

Every year, many rural Kenyans move to the cities to find work in manufacturing or service industries. Urban factory workers help manufacture such products as cement, chemicals, household utensils, light machinery, motor vehicles, paper, processed food, and textiles. Other urban Kenyans work in stores and offices.

Most city people live in modern houses made of stone or cement. Some are simple working-class homes; others are large, expensive houses or apartments for the well-to-do.

About 40,000 Kenyans work in the tourism industry. Every year, more than 350,000 tourists visit Kenya to enjoy the tropical coast and, more often, to view and photograph Kenya's magnificent wild animals. These tourists spend more than $200 million in Kenya, contributing more to the nation's economy than any other single activity except coffee growing.

Kiribati

The small island country of Kiribati (pronounced *KIHR ih bahs*) in the southwest Pacific Ocean is made up of the 16 Gilbert Islands, Ocean Island (Banaba), the 8 Phoenix Islands, and 8 of the Line Islands. Although it has a total land area of only 277 square miles (717 square kilometers), its islands are scattered over about 2 million square miles (5 million square kilometers) of the Pacific.

Almost all of Kiribati's islands are coral reefs, and many are *atolls,* ring-shaped reefs that enclose a lagoon. Kiribati has a tropical climate, with temperatures of about 80° F. (27° C) the year around. The annual rainfall is about 100 inches (254 centimeters) in the northern islands, and about 40 inches (100 centimeters) in the other islands.

People and history

Most of Kiribati's people—called *I-Kiribati*—are Micronesians, and about 96 per cent live in the Gilbert Islands. Tarawa, one of the Gilberts, is the country's capital. Tarawa is an atoll composed of many coral islets.

Most I-Kiribati live in rural villages of 10 to 170 houses clustered around a church and a meeting house. Many of the village houses are made of wood and leaves from palm

trees. The language of the islanders is Gilbertese, but most people also speak some English. The people of Kiribati grow most of their own food, which includes bananas, breadfruit, papaya, and giant taro. The islanders also raise pigs and chickens and catch fish for food. They wear light cotton clothing.

Most of the I-Kiribati are descended from Samoans who invaded the islands about 1400 and from people who had settled there earlier. In the 1500's, Spanish explorers became the first Europeans to sight the islands.

During the 1890's, the United Kingdom took control of the Gilbert Islands and other neighboring islands. The United Kingdom gained control of Ocean Island in 1901. It made all these islands a colony in 1916. Some of the Line Islands and the Phoenix Islands were added to the colony later.

During World War II (1939–1945), Japanese troops occupied several of the islands. The United States Marines invaded Tarawa in 1943 and defeated the Japanese in one of the bloodiest battles of the war. Thirty-three islands in the former colony gained independence as Kiribati on July 12, 1979.

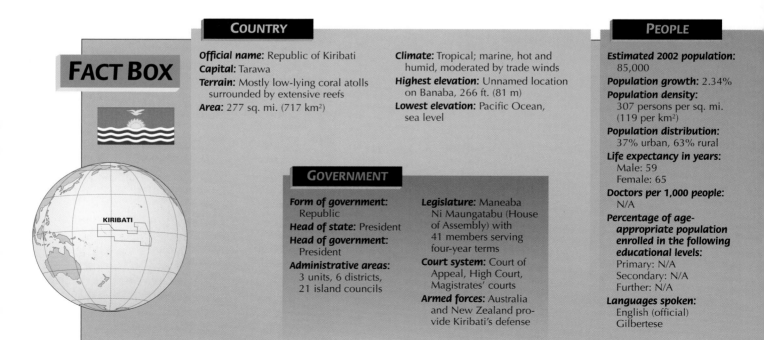

FACT BOX

COUNTRY

Official name: Republic of Kiribati
Capital: Tarawa
Terrain: Mostly low-lying coral atolls surrounded by extensive reefs
Area: 277 sq. mi. (717 km²)

Climate: Tropical; marine, hot and humid, moderated by trade winds
Highest elevation: Unnamed location on Banaba, 266 ft. (81 m)
Lowest elevation: Pacific Ocean, sea level

GOVERNMENT

Form of government: Republic
Head of state: President
Head of government: President
Administrative areas: 3 units, 6 districts, 21 island councils

Legislature: Maneaba Ni Maungatabu (House of Assembly) with 41 members serving four-year terms
Court system: Court of Appeal, High Court, Magistrates' courts
Armed forces: Australia and New Zealand provide Kiribati's defense

PEOPLE

Estimated 2002 population: 85,000
Population growth: 2.34%
Population density: 307 persons per sq. mi. (119 per km²)
Population distribution: 37% urban, 63% rural
Life expectancy in years:
Male: 59
Female: 65
Doctors per 1,000 people: N/A
Percentage of age-appropriate population enrolled in the following educational levels:
Primary: N/A
Secondary: N/A
Further: N/A
Languages spoken:
English (official)
Gilbertese

Traditional dress on the Gilbert Islands included shell and bone jewelry and a grass skirt, as shown in this turn-of-the-century photograph. Gilbert Islanders are now citizens of the nation of Kiribati. About 96 per cent of Kiribati's people live in the Gilberts.

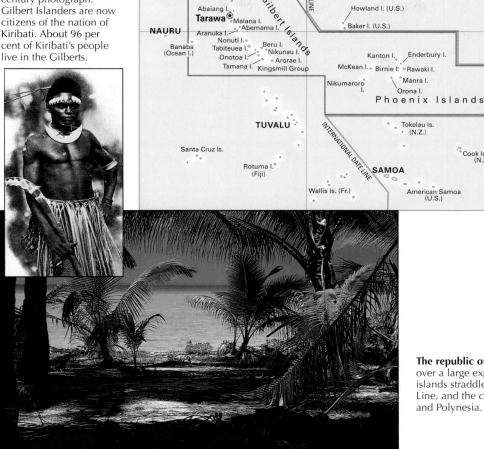

The republic of Kiribati consists of 33 islands spread over a large expanse of the Pacific, *map above.* The islands straddle the equator, the International Date Line, and the cultural boundary between Micronesia and Polynesia.

ECONOMY

Religions:
Roman Catholic 53%
Protestant (Congregational) 41%
Seventh-Day Adventist
Baha'i
Church of God
Mormon

TECHNOLOGY

Radios per 1,000 people: N/A

Televisions per 1,000 people: N/A

Computers per 1,000 people: N/A

Currency: Australian dollar
Gross national income (GNI) in 2000: $86 million U.S.
Real annual growth rate (1999–2000): 1.8%
GNI per capita (2000): $950 U.S.
Balance of payments (2000): N/A
Goods exported: Mostly: copra Also: seaweed, fish
Goods imported: Foodstuffs, machinery and equipment, miscellaneous manufactured goods, fuel
Trading partners: Australia, United States, New Zealand, Fiji, Japan

Coconut palms frame a coral lagoon in Kiribati. Coconuts furnish a livelihood for many of the country's island communities. Copra is Kiribati's only important export. Bananas, breadfruit, papaya, and giant taro also grow on the islands.

Government and economy

Kiribati is a republic headed by a president. The people elect the president from among candidates nominated by the House of Assembly. The House, the nation's lawmaking body, consists of 41 members whom the people elect to four-year terms. Most of Kiribati's inhabited islands have a local governing council. The government receives economic aid from Australia, the United Kingdom, and New Zealand.

Kiribati's only important export is copra. The commercial and shipping center of the country is Betio, a densely populated islet in southwest Tarawa. Bairiki, east of Betio, is the government center. Bonriki, in the southeast, has an international airport.

Korea, North

Kim Il Sung was the leader of the Democratic People's Republic of Korea—usually called North Korea—from the time his Communist government was established in 1948 until his death in 1994. Kim's government was a rigid dictatorship and taught people that Kim was the "sun" of all the people and could do no wrong. Kim's son and designated heir, Kim Jong II, succeeded his father as head of state.

After Kim Il Sung came to power, his government took farmland from wealthy landowners and gave it to the farmworkers. Then in the 1950's, the many small farms were made into large government-controlled collectives. The government also took over most of North Korea's industries, emphasizing the development of heavy industry and military power.

In the 1950's, the country made impressive gains in achieving industrialization, but its goal of economic self-sufficiency restricted its progress.

Foreign relations

Between 1948 and 1971, North Korea was one of the most isolated countries in the world. During this period, the country's foreign relations were conducted almost exclusively with other Communist nations, particularly China and the Soviet Union.

During the 1970's, however, military and economic assistance from both countries decreased. North Korea turned to Third World countries for trade and cultural exchanges, but these negotiations were largely unsuccessful.

When the Soviet Union and China began increasing their contacts with South Korea in the 1980's, President Kim, fearing total isolation, tried to improve the nation's relations with other countries. In 1988, he announced his intention to "develop economic and technical cooperation and cultural exchange" with countries that did not have formal relations with North Korea. Kim also repeated his long-standing proposal that North Korea and South Korea unite to form a single nation combining Communist and capitalist forms of government.

In 2002, North Korea held talks with Japan, the United States, and other countries in efforts to establish more friendly relations. However, these efforts were thwarted when North Korea revealed that it had a secret program to develop nuclear weapons. The program violated North Korea's 1994 agreement with the United States promising to halt all nuclear weapons-related activities in return for energy assistance. In late 2002 and early 2003, North Korea expelled international atomic energy inspectors from the country and reactivated its nuclear facilities.

FACT BOX

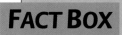

NORTH KOREA

COUNTRY

Official name: Choson-minjujuui-inmin-konghwaguk (Democratic People's Republic of Korea)
Capital: P'yongyang
Terrain: Mostly hills and mountains separated by deep, narrow valleys; coastal plains wide in west, discontinuous in east
Area: 46,541 sq. mi. (120,540 km²)
Climate: Temperate, with rainfall concentrated in summer
Main rivers: Yalu, Taedong, Tumen
Highest elevation: Paektu-san, 9,003 ft. (2,744 m)
Lowest elevation: Sea of Japan, sea level

GOVERNMENT

Form of government: Authoritarian socialist
Head of state: Chairman of the National Defense Commission
Head of government: Premier
Administrative areas: 9 do (provinces), 3 si (special cities)
Legislature: Ch'oego Inmin Hoeui (Supreme People's Assembly) with 687 members serving five-year terms
Court system: Central Court
Armed forces: 1,082,000 troops

PEOPLE

Estimated 2002 population: 24,585,000
Population growth: 1.35%
Population density: 528 persons per sq. mi. (204 per km²)
Population distribution: 59% urban, 41% rural
Life expectancy in years:
Male: 68
Female: 74
Doctors per 1,000 people: 3.0
Percentage of age-appropriate population enrolled in the following educational levels:
Primary: N/A
Secondary: N/A
Further: N/A

CHINA

The Democratic People's Republic of Korea occupies the northern part of the Korean Peninsula. Visitors, who were allowed in during one week in 1989, reported that the capital, Pyongyang, was a bleak city in which the basic necessities of life were rationed.

A parade in North Korea is generally a display of the country's military strength. North Korea's economic development suffered as a result of the government's determination to build up its armed forces.

ECONOMY

Language spoken: Korean

Religions:
Buddhism
Confucianism
Christian
Chondogyo (Religion of the Heavenly Way)

TECHNOLOGY

Radios per 1,000 people: 154

Televisions per 1,000 people: 54

Computers per 1,000 people: N/A

Currency: North Korean won

Gross domestic product (GDP) in 1999: $22.6 billion U.S.

Real annual growth rate (1999): 1%

GDP per capita (1999): $1,000 U.S.

Balance of payments (2000): N/A

Goods exported: Minerals, metallurgical products, manufactures (including armaments), agricultural and fishery products

Goods imported: Petroleum, coking coal, machinery and equipment, consumer goods, grain

Trading partners: Japan, China, South Korea, Russia

Relations with South Korea

Hostilities between North Korea and South Korea have existed since 1945. During the 1960's and 1970's, North Korea's frequent attacks in the demilitarized zone (DMZ) created worldwide tension, and each country has claimed to be the only legitimate government on the Korean Peninsula.

In December 1991, North and South Korea signed a treaty promising to end aggression and permit travel, trade, and humanitarian exchanges between the two countries. In 2000, the leaders of North and South Korea met to discuss relations. As a result of this meeting, the two countries have made some additional moves toward improving relations. For example, some North Korean and South Korean relatives have been allowed to visit one another, and a road has been opened between the two countries.

Land and People

North Korea's three main land regions are the Northwestern Plain, the Northern Mountains, and the Eastern Coastal Lowland. The Northwestern Plain stretches along the entire western coast of North Korea, divided by rolling hills into a series of broad, level plains. The plains have most of the country's farmland, as well as its major industrial areas, including Pyongyang. About 50 per cent of North Korea's people live in the region.

The Northern Mountains region, east of the Northwestern Plain, covers almost all of central North Korea. The forested mountains contain rich mineral deposits of coal, copper, gold, graphite, iron ore, lead, magnesium, silver, tin, tungsten, and zinc.

Paektu-san (Paektu Mountain), the highest mountain on the Korean Peninsula, rises 9,003 feet (2,744 meters) in the Northern Mountains on the border between North Korea and China. The Yalu River, the longest river in North Korea, flows westward from this mountain along the border to the Yellow Sea. The Tumen River forms the border eastward from Paektu-san to the Sea of Japan. About 25 per cent of North Korea's people live in this region.

The Eastern Coastal Lowland, a strip of land between the Northern Mountains and the Sea of Japan, covers most of the country's east coast. A series of narrow plains separated by low hills provide excellent farmland, and the region is an important agricultural, fishing, and industrial area. More than 25 per cent of North Korea's people live in the Eastern Coastal Lowland.

Climate

North Korea's climate is affected by monsoons throughout the year. A summer monsoon brings hot, humid weather, with temperatures in July and August averaging about 85° F. (30° C) in Pyongyang. Mountains protect the eastern plains from the cold, dry winter monsoon, producing mild winter seasons, but winters on the northern border are very cold, with temperatures in January averaging −5° F. (−20° C). Annual precipitation in North Korea generally ranges from 30 to 60 inches (76 to 150

North Korean children are required to attend school for 11 years, including a year of pre-school. All children are brought up in public nurseries and kindergartens until they are 5 years old. Parents are allowed to bring them home only once a week.

Visitors enjoy the view from a bridge in scenic Myohyang Mountains in North Korea's Northern Mountains region. Forested mountains and hills cover most of the country.

Workers prepare for the rice harvest, *left,* on a North Korean collective farm. The workers receive a share of the products and some cash payment in exchange for their labor.

Construction workers at a dam site take a break, *above.* All industry in North Korea is owned by the government, and workers are compensated according to the quality and quantity of work done.

centimeters), but most of the country's rain falls in the summer months between June and August.

Life under Communism

After the Korean War, North Korea's Communist government made many changes in the traditional Korean way of life. The government's emphasis on industrialization encouraged many North Koreans to leave rural farm areas for jobs in the cities, thus weakening strong family ties. In addition, the Communists actively discouraged religion because they felt it conflicted with the teachings of Communism.

A government-designed program made an idol of President Kim Il Sung, to replace the North Korean people's traditional Confucian values and family loyalties. The North Korean media constantly acclaimed Kim's genius as a leader, and the people were required to spend at least two hours a day studying his writings. By idolizing the president, the government hoped to inspire the people's unquestioned devotion to Kim's policies.

One such policy, established in North Korea's Constitution, is the requirement that all adults work. Women hold equal status with men in North Korean society and are employed in every industrial and agricultural sector, including coal mining.

Industry employs approximately 30 per cent of the country's workers, most of whom live in cities and towns in large, state-built and state-owned apartment buildings. About 45 per cent of the North Korean people work in agriculture and live on state-controlled collective farms. The country has about 3,700 collectives, with about 300 families living on each one.

North Korea's Communist regime also controls the cultural and social life of the people. The government provides recreation and vacation facilities for the workers and dictates the type of entertainment available. Art forms that conflict with Communist principles are prohibited.

Korea, South

The Republic of Korea—usually called South Korea—covers the southern half of the Korean Peninsula in eastern Asia. The Communist-ruled Democratic People's Republic of Korea—usually called North Korea—lies to the north. The Korean Peninsula is bounded by the Sea of Japan on the east, the Yellow Sea on the west, and the Korea Strait on the south. The northern part of the peninsula is bordered mostly by China and a small area of Russia.

Seoul, South Korea's capital and largest city, is also the country's cultural, economic, and educational center. Nearly 25 per cent of South Korea's population, about 10 million people, live in Seoul.

Mountains extend through most of central and eastern South Korea. The graceful Buddhist shrine at Hajodae perches along the rocky eastern coastline of the Taebaek Mountains, which plunge eastward into the Sea of Japan. Plains cover much of the rest of the country, and numerous islands lie off the southern and western coasts of the peninsula.

Korea was a single nation before World War II (1939-1945). The two separate states were formed in 1948, and although the two countries have pursued very different political courses since then, they share a long history.

Early history

People lived in the southwestern part of the Korean Peninsula about 30,000 years ago, and the earliest Korean state, called Choson, developed in the northern part of the peninsula many centuries before the birth of Christ. In 108 B.C., China conquered parts of this territory and strongly influenced Korean arts, sciences, and government. The Chinese also introduced Buddhism to Korea.

During the A.D. 100's, the kingdom of Koguryo emerged in northeastern Korea, and two other Korean states—Paekche in the southwest and Silla in the southeast—were formed in the 200's. Silla gained control of the other two kingdoms and united the peninsula in the 660's. Art and learning flourished in Silla, and Confucianism was introduced from China. Provincial warlords broke up the kingdom in the 800's, but Silla was reunited in 932 and renamed

Koryo. The word *Korea* came from the word *Koryo*.

In 1388, a general named Yi Songgye took control of Koryo, and in 1392 he became king and renamed the country Choson. In the 1630's, Manchu armies from the north defeated the Koreans, and though members of the Yi family continued as kings, the Manchus demanded payments from Korea until the late 1800's.

Between the 1600's and 1800's, Korea was called the *Hermit Kingdom* because the country was closed to foreigners for almost 200 years. In 1876, however, Japan forced the kingdom to open some ports to trade, and Japan's influence in Korea became stronger than that of China. Japan took complete control in 1910 and ruled Korea until Japan's defeat in World War II (1939-1945).

The Korean War

When World War II ended, Soviet troops occupied the northern half of the Korean Peninsula, and U.S. troops occupied the southern half. In 1948, two separate Korean states were formed — the Democratic People's Republic of Korea in the north and the Republic of Korea in the south, and by 1949 both the Soviet Union and the United States had withdrawn their troops from the peninsula.

In June 1950, troops from Communist-ruled North Korea invaded South Korea, and the Korean War began. The United Nations (UN) demanded that the Communists withdraw, and when the Communists refused to comply, the UN called on its member nations to give military aid to South Korea. Sixteen countries sent troops and 41 countries sent military supplies, but the United States sent more than 90 per cent of the troops and supplies. China fought on the side of North Korea, and the Soviet Union gave the Communists military aid.

The Korean War ended on July 27, 1953. An armistice agreement established a buffer zone called the Demilitarized Zone (DMZ) between North and South Korea, but a permanent peace treaty has never been signed.

South Korea Today

The Korean War had destroyed many of South Korea's factories and ruined farm crops, so the nation was faced with grave economic problems when the fighting stopped. President Syngman Rhee's efforts to develop his country's economy were based largely on aid from the United States, and he failed to develop an economic plan for South Korea. As a result, during the early 1950's, members of the National Assembly, the country's legislature, became increasingly critical of Rhee.

Rhee's growing use of undemocratic methods to keep control of the government eventually aroused opposition among the South Korean people, especially the students. In 1960, Rhee and his party fixed an election to ensure victory for his vice presidential candidate. In protest, students led widespread demonstrations against the government, and Rhee was forced to resign.

The new government was weak, however, and in 1961 it was overthrown by a group of military officers led by General Park Chung Hee. In 1963, Park was elected president of South Korea, and his Democratic Republican Party won a majority of the seats in the National Assembly.

South Korea's economy developed rapidly under Park, but he used his power to hold down opposition and deny the people's civil rights. Freedom of speech and freedom of the press were limited, and many of Park's political opponents were jailed. In 1972, he put a new constitution in force that gave him almost unlimited powers. It provided that the president would be chosen by members of the Electoral College—rather than elected directly by the people—and could serve an unlimited number of terms.

In 1979, Park was assassinated. The country's Electoral College elected an interim president, but the government delayed a promised constitutional revision that would allow for direct presidential elections. Many South Koreans took part in demonstrations that were brutally put down by the government, and in 1980 martial law was declared. In August of that year, Lieutenant General Chun Doo Hwan was elected president by the Electoral College.

Throughout most of the 1980's, students demonstrated against Chun, demanding a new, more democratic constitution, while a few rebels voiced support for North Korea's Communist government. The South Korean government used the actions of this Communist minority to justify harsh measures in dealing with public dissent.

In 1987, mounting public sympathy for the demonstrators finally forced Chun to allow direct election of the president. The election

FACT BOX

COUNTRY

Official name: Taehan-min'guk (Republic of Korea)
Capital: Seoul
Terrain: Mostly hills and mountains; wide coastal plains in west and south
Area: 38,023 sq. mi. (98,480 km²)

Climate: Temperate, with rainfall heavier in summer than winter
Main rivers: Naktong, Han
Highest elevation: Halla-san, 6,398 ft. (1,950 m)
Lowest elevation: Sea of Japan, sea level

GOVERNMENT

Form of government: Republic
Head of state: President
Head of government: Prime minister
Administrative areas: 9 do (provinces), 6 gwangyoksi (special cities)

Legislature: Kukhoe A(National Assembly) with 273 members serving four-year terms
Court system: Supreme Court
Armed forces: 672,000 troops

PEOPLE

Estimated 2002 population: 47,521,000
Population growth: 0.93%
Population density: 1,240 persons per sq. mi. (479 per km²)
Population distribution: 79% urban, 21% rural
Life expectancy in years: Male: 71 Female: 79
Doctors per 1,000 people: 1.3
Percentage of age-appropriate population enrolled in the following educational levels: Primary: 94 Secondary: 102* Further: 68
Languages spoken: Korean English

was held in December 1987, and Roh Tae Woo was elected president.

Roh promised to give the South Koreans a government they could trust. He also made a number of proposals during 1988 to improve relations with North Korea, calling for trade between North Korea and South Korea and authorizing family visits and student exchanges.

From 1945 until the early 1990's, the relationship between North Korea and South Korea was one of mutual suspicion, and American troops have guarded the neutral demilitarized zone since 1953. However, in December 1991, North and South Korea signed a treaty ending aggression and permitting travel, trade, and humanitarian exchanges between the two countries. In 2002, the two countries made some additional moves toward improving relations. A road has been opened between the two countries and some North Korean and South Korean relatives have been allowed to visit one another.

Seoul's South Gate dates from the city's founding in the late 1300's, *top right.* Although South Koreans feared terrorist attacks by North Korea during the 1988 Summer Olympic Games, peace prevailed, and the Summer Games were a great success.

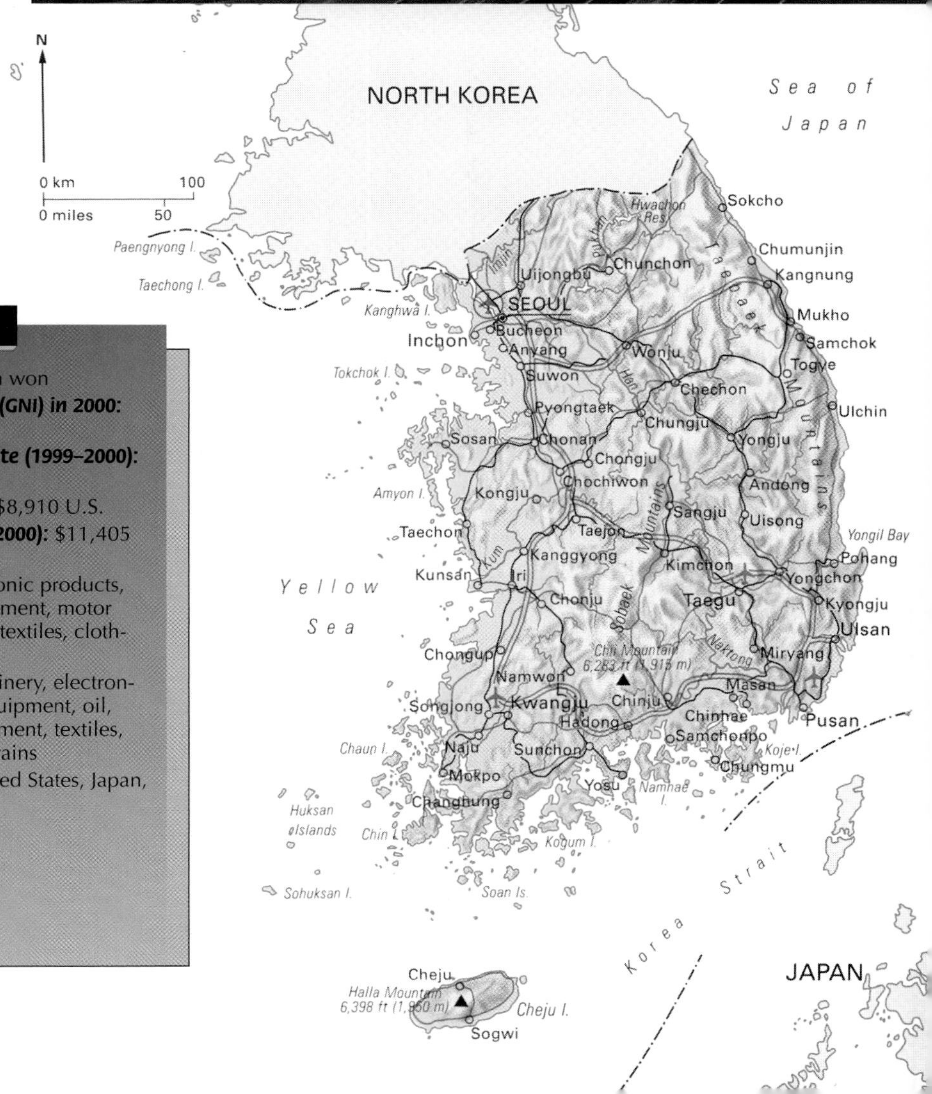

The Republic of Korea, or South Korea, has been divided from North Korea since 1945. This division continues to be a source of bitter sorrow for many Koreans.

Religions: Christian 49%
Buddhist 47%
Confucianist 3%
Shamanist
Chondogyo (Religion of the Heavenly Way)

Enrollment ratios compare the number of students enrolled to the population which, by age, should be enrolled. A ratio higher than 100 indicates that students older or younger than the typical age range are also enrolled.

TECHNOLOGY

Radios per 1,000 people: 1,033
Televisions per 1,000 people: 364
Computers per 1,000 people: 237.9

ECONOMY

Currency: South Korean won
Gross national income (GNI) in 2000: $421.1 billion U.S.
Real annual growth rate (1999–2000): 8.8%
GNI per capita (2000): $8,910 U.S.
Balance of payments (2000): $11,405 million U.S.
Goods exported: Electronic products, machinery and equipment, motor vehicles, steel, ships, textiles, clothing, footwear, fish
Goods imported: Machinery, electronics and electronic equipment, oil, steel, transport equipment, textiles, organic chemicals, grains
Trading partners: United States, Japan, China

Land and People

South Korea has three main land regions—the Central Mountains, the Southern Plain, and the Southwestern Plain. The Central Mountains region extends throughout most of central and eastern South Korea and into a small part of southern North Korea. Lush forests cover the mountains in this region, even extending along much of the seacoast. Hillsides, river valleys, and some land along the eastern coast are used for farming, while coastal waters yield plentiful supplies of fish. More than 25 per cent of South Korea's people live in the Central Mountains region.

The Southern Plain, which covers the country's entire southern coast, consists of a series of plains separated by low hills. Pusan, one of the country's major industrial centers, is located in the Southern Plain, and South Korea's longest river, the Naktong, flows through the region from mountains in the north to the Korea Strait. Nearly 25 per cent of South Korea's people live in this important agricultural and industrial region.

The Southwestern Plain, which extends along almost the entire western coast of South Korea, is covered by rolling hills and plains and has much of the country's best farmland as well as the important industrial area around Seoul. The Han River flows through the region from mountains in the east to the Yellow Sea. About 50 per cent of South Korea's people live in the Southwestern Plain region.

About 3,000 islands lie off the southern and western coasts of the Korean Peninsula. However, only about 200 of these islands are inhabited. The largest island, Cheju, lies about 50 miles (80 kilometers) south of the peninsula and covers about 700 square miles (1,800 square kilometers). South Korea's highest mountain, Halla-san (Halla Mountain) towers 6,398 feet (1,950 meters) on this island. Cheju has its own provincial government, while other islands are governed by provinces on the mainland.

South Korea's climate is affected the year around by monsoon winds. During the summer, a monsoon blows in from the south and southeast, bringing hot, humid weather and temperatures averaging between 70° F. (21° C) and 80° F. (27° C). In the winter, a cold, dry monsoon blowing in from the north and northwest brings cold weather. January temperatures in southeastern Korea average about 35° F. (2° C).

Most of South Korea receives from 30 to 50 inches (76 to 130 centimeters) of *precipitation* (rain, melted snow, and other forms of moisture) per year, with heavy rainfall from June through August accounting for about half of that total. Strong tropical storms called *typhoons* pose a threat to the Korean Peninsula during July and August.

The population of the Korean Peninsula consists almost entirely of Koreans—descendants of the people who settled in the region more than 30,000 years ago. Most Koreans are small in stature, with broad faces, straight black hair, olive-brown skin, and dark eyes. People of Chinese descent make up the largest minority group on the peninsula.

South Korea has a population of about 44 million people, with more than 72 per cent living in urban areas. Rural people began moving to the cities after Japan seized control of Korea in 1910 and brought industry to these areas. The move to the cities increased after World War II (1939-1945) and the Korean War (1950-1953), and during the 1960's and 1970's thou-

An elderly husband wears a tall, black horsehair hat, a traditional sign of his married state. Married Korean men also customarily wore their hair tied in a topknot.

Sorak-san (Snow Peak Mountain) National Park, *left,* located on South Korea's eastern coast in the Central Mountains region, has some of the country's most striking scenery.

A fishing village on South Korea's southern coast, *above,* overlooks rocky islands off the mainland. Many South Koreans make their living by fishing, and seafood forms an important part of the people's diet.

Bicycles crowd busy Yoido Plaza in Seoul, *center.* Many people live in high-rise apartment buildings on Yoido Island in the Han River, an area of the city just southwest of the downtown district.

Outdoor dining is a popular activity in Seoul's South Gate market, where the favorite dishes include *kimchi,* a highly spiced mixture of Chinese cabbage, onions, white radishes, and other vegetables.

sands of people moved to Seoul and other cities to work in the factories being built in these centers.

The traditional Korean way of life began to deteriorate as a result of this urbanization. Family ties in South Korea weakened as young people moved to the cities, and South Koreans were also exposed to Western influences through trade and political relations with Western nations. Many people, particularly in the cities, wear Western-style clothing, and numerous skyscrapers and high-rise apartment buildings have been erected in urban areas.

Traditions have generally remained stronger in South Korea's rural areas, however. For example, colorful traditional dress is still worn by many people in the countryside, particularly by women. In addition, most housing in rural areas consists of traditional one-story houses heated by a network of pipes under the floor. This heating system, called *ondol,* is so much a part of Korean life that many modern urban homes have ondol floors.

791

Economy

South Korea enjoys one of the world's fastest-growing industrial economies. The boom began in the 1960's when President Park Chung Hee focused South Korea's economy on manufacturing and export trade, rather than agriculture.

To achieve its goals, the government influenced and sometimes controlled industrial firms and financial institutions. For example, companies that met their export targets were rewarded, while firms that failed could lose bank credit. Foreign aid, particularly from the United States and Japan, also helped Park carry out his program of economic restructuring.

In the early 1960's, economic growth in South Korea came mainly from light industry, such as textiles. By keeping production costs low, the goods could be sold at highly competitive prices on the world market. By the 1970's and 1980's, however, the nation's heavy, defense, and chemical industries were developing rapidly, and South Korea continued its successful export drive.

Fish merchants bid for the day's catch at an auction in Pusan, South Korea's major port and an important center for the country's fishing fleet.

Planting rice by hand in flooded paddy fields, *right*, is a slow and back-breaking job. Today, many South Korean farms use modern agricultural methods, and machines do much of the work.

Industry

Today, manufacturing accounts for about 75 per cent of South Korea's industrial production. The nation's fast-growing electronics and automobile industries manufacture goods that rival those produced by the world's leading industrial nations, and South Korea also ranks as a major producer of chemicals, fertilizers, iron and steel, machinery, and ships. Other manufactured products include electrical appliances, optical goods, paper, plywood, porcelain, and rubber tires.

About 25 per cent of all South Koreans work in industry. Food processing and the manufacture of clothing, shoes, and textiles employ most of these workers. About 50 per cent of the people work in service activities.

South Korea's change from an agricultural to an industrial society also spurred a boom in construction. Factories, office and apartment buildings, highways, and water and sewer systems have been built throughout the country. Today, construction accounts for about 20 per cent of South Korea's industrial production, and mining

The massive Hyundai shipyard at Ulsan on South Korea's east coast is owned by one of the country's powerful conglomerates. Developed in the 1960's, these large companies helped carry out the government's industrialization and export drive.

High-technology manufacturing of microchips, computers, and other sophisticated products reflects the growing complexity of South Korea's economy, *above*.

accounts for about 5 per cent, with *anthracite* (hard coal) and tungsten as the chief mining products.

Agriculture and fishing

The nation's agricultural development was slower than that in industry, but South Korea enjoyed its most prosperous agricultural period in the 1970's due to modern farming methods and government incentives. As a result, in the 1980's, the country achieved its goal of becoming self-sufficient in rice production.

Rice is by far South Korea's chief crop. South Korean farms also produce apples, barley, Chinese cabbage, onions, potatoes, sweet potatoes, white radishes, hogs, chickens, and eggs. Oranges are grown on the island of Cheju, off Korea's southern coast.

The number of people employed in agriculture has decreased with industrialization. While more than 50 per cent of South Korea's population worked on farms in the 1960's, agriculture employs only about 25 per cent of the people today. Mechanization has helped solve the labor shortage in rural areas.

The modernization of its fishing industry has made South Korea one of the world's leading fishing countries. The catch includes filefish, oysters, and pollack. Many farmers supplement their income by fishing.

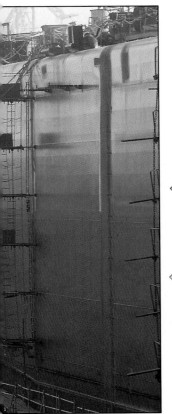

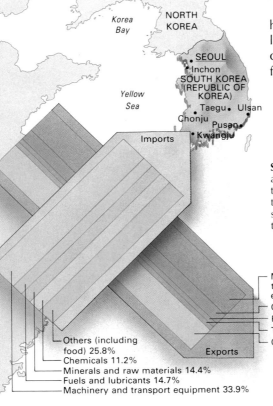

South Korea's economic growth after the vast devastation caused by the Korean War stands as one of the most remarkable success stories in international development. Today, with few natural resources, the country imports the materials needed to manufacture finished export goods.

Machinery and transport equipment 35.8%
Clothing 15.9%
Footwear 5.8%
Textiles 5.1%
Others 37.4%

Others (including food) 25.8%
Chemicals 11.2%
Minerals and raw materials 14.4%
Fuels and lubricants 14.7%
Machinery and transport equipment 33.9%

Imports

Exports

Korea Bay

NORTH KOREA

SEOUL
Inchon
SOUTH KOREA
(REPUBLIC OF KOREA)

Yellow Sea

Taegu Ulsan
Chonju
Pusan
Kwangju

Culture and Art

Korea's oldest faith—Shamanism—was brought to the peninsula more than 30,000 years ago by the people who first settled in the region. Shamanism is based on the belief that spirits reside in natural forces and nonliving objects, such as thunder and rocks, as well as in human beings and animals. According to this faith, the shaman, a priestlike figure, can communicate with these spirits and pacify them by performing certain rituals. Shamanism persists today in some parts of South Korea.

Following the formation of the Three Kingdoms—Koguryo, Paekche, and Silla—in the late A.D. 200's, Taoism, Buddhism, and Confucianism were introduced into Korea from China. The Taoist emphasis on living in harmony with nature is reflected on the South Korean flag in a Taoist symbol called *yin* and *yang* that represents the balance in the universe between opposites—such as male and female, and life and death.

Buddhism became the chief religion of the Three Kingdoms during the 300's and 400's, and temples and monasteries constructed during this time became centers of learning and of the arts. Much of Korean art reflects the influence of Buddhist teaching, and Buddha's birthday is a national holiday in South Korea.

Confucianism, which came to Korea after Silla absorbed the other two kingdoms in the 660's, has traditionally been the most widely followed set of beliefs. Confucianism stresses the duties that people have toward one another, particularly within the family. Today, most South Korean families continue to honor their ancestors in special ceremonies, according to the teachings of Confucius.

Roman Catholic missionaries brought Christianity to Korea during the 1830's. Christianity influenced modern Korea through its emphasis on learning and social reform.

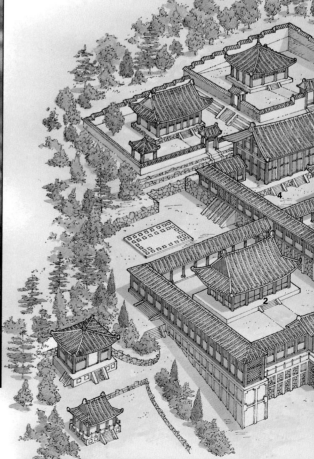

Drummers pound out a rhythm on the *changgo* at a traditional Korean farmers' dance. At these spirited affairs, some dancers play the changgo, while others play smaller drums, gongs, trumpets, or oboes. One performer twirls a long white streamer attached to his hat and leaps in and out of its circles.

A dragon adorning the Sangwonsa Temple in Mount Odae National Park, *below,* reflects the influence of Shamanism. According to this faith, the dragon, tiger, turtle, and phoenix guard the universe on the west, east, north, and south.

Honoring their royal ancestors, descendants of Korea's Yi dynasty, which ruled from 1392 until 1910, solemnly take part in a Confucian ceremony at Chongmyo Shrine, *far left.*

Pulguk-sa Temple near Kyongju is Korea's oldest Buddhist shrine. Built in the A.D. 500's, the splendors of this ancient temple include ornate gateways **(1)**, the golden Buddha of Kuknak-jon Hall **(2)**, the hall of worship **(3)**, and the meeting hall **(4)**.

Music and dance

Traditional Korean court music developed largely under Chinese influence. Although Korea's last royal dynasty ended in 1910 when the Japanese took control of the country, the court music and its ancient instruments were preserved in South Korea. Today, court orchestras perform the music throughout the nation and keep this ancient heritage alive.

Folk music is another important element of traditional Korean music. Traditional Korean folk musicians used a variety of special instruments, including an hourglass drum called a *changgo* and a 12-stringed zither called a *kayagum.*

Folk music and folk dance were used primarily to express religious beliefs. Shamanist-inspired farmers' dances, for example, were performed in the spring to ask for good crops, and in the fall to give thanks for the harvest. Folk songs expressing Buddhist or Confucian beliefs celebrated the virtues of honoring parents and chastity.

Masked dance dramas, on the other hand, were intended to entertain audiences. Accompanied by music and song, these dramas used wit to make fun of the upper class and life in general.

Poetry, painting, and ceramics

Until the 1400's, Koreans used Chinese characters to express their thoughts in written form, having no writing system of their own. Then they developed their own alphabet, called *hangul,* with 24 characters. Korean poetry, traditionally the most admired form of writing, often has a religious theme, but nature has also served as an inspiration in both Korean literature and painting.

Art in Korea was closely associated with Confucian scholarship, which emphasized poetry, *calligraphy* (the art of beautiful handwriting), and painting. Korean art generally included landscapes, animals, and the "four gentlemen" — bamboo, chrysanthemum, orchid, and plum blossoms.

However, many historians believe that Korean art reached its greatest heights in ceramics. The blue-green pottery known as *celadon* is particularly admired for its graceful lines and depth of color.

Kuwait

Kuwait is a tiny Arab nation at the northern end of the Persian Gulf. A poor desert land until 1946, Kuwait became one of the richest and most progressive countries of the world when vast petroleum deposits were discovered under its deserts.

Kuwait's rulers used its oil profits to turn a desert wilderness into a prosperous welfare state. Now one of the world's wealthiest nations in terms of national income per person, Kuwait has free education, free health and social services—and no income tax.

The population of Kuwait is 28 times larger than it was in the 1930's, due mainly to immigration, and less than half of the people are Kuwaiti citizens. Most of the other residents of Kuwait are Arabs from foreign lands, Asian Indians, Iranians, or Pakistanis. Palestinian Arabs are by far the largest group of non-Kuwaiti residents in the country. Most of the people are Arabs and Muslims.

Kuwait is a hot, dry land. An average of 2 to 6 inches (5 to 15 centimeters) of rain falls each year from October to March. Some grass may grow then, but otherwise Kuwait has little vegetation except desert scrub. The country has no rivers or lakes, and before 1950 drinking water was brought by ships from Iraq. Today, fresh water is produced in Kuwait by distilling seawater and mixing it

with well water. In addition, a large underground source of fresh water has been discovered.

The region that is now Kuwait had few settlements until the 1700's, when members of an Arab tribe found fresh water on the southern shore of Kuwait Bay and settled in what is now the port city of Kuwait. The group elected the head of the Al-Sabah family to rule them as Sabah I, and descendants of this family, headed by a powerful leader called an *emir,* ruled Kuwait until August 1990, when Iraqi forces invaded and occupied the country. Iraq's president, Saddam Hussein, justified the action by claiming that Kuwait was legally a part of Iraq.

The United Nations (UN) condemned the invasion of Kuwait and refused to recognize Hussein's declaration that his country had annexed Kuwait as an Iraqi province. On Jan. 17, 1991, after months of pressuring Iraq to leave Kuwait, a coalition of international forces led by the United States began conducting bombing attacks on Iraqi military and industrial targets.

In late February, the Allied forces launched a massive ground attack into Kuwait and southern Iraq and quickly defeated the Iraqi troops. On Feb. 26, Hussein ordered his troops to withdraw from Kuwait, thus freeing

FACT BOX

COUNTRY

Official name: Dawlat al Kuwayt (State of Kuwait)
Capital: Kuwait
Terrain: Flat to slightly undulating desert plain
Area: 6,880 sq. mi. (17,820 km²)

Climate: Dry desert; intensely hot summers, short, cool winters
Highest elevation: Unnamed location, 928 ft. (283 m)
Lowest elevation: Persian Gulf, sea level

KUWAIT

GOVERNMENT

Form of government: Nominal constitutional monarchy
Head of state: Emir
Head of government: Prime minister
Administrative areas: 5 muhafazat (governorates)

Legislature: Majlis al-Umma (National Assembly) with 50 members serving four-year terms
Court system: High Court of Appeal
Armed forces: 15,300 troops

PEOPLE

Estimated 2002 population: 2,062,000
Population growth: 3.44%
Population density: 300 persons per sq. mi. (116 per km²)
Population distribution: 100% urban
Life expectancy in years:
Male: 75
Female: 77
Doctors per 1,000 people: 1.9
Percentage of age-appropriate population enrolled in the following educational levels:
Primary: 77
Secondary: 65
Further: 19

The blend of traditional and modern in Kuwait is represented by a minaret standing beside a relay tower of the national radio station. From the minaret, a muezzin calls Muslims to prayer, while the tower beams a variety of radio programs.

Crown Prince Sheik Saad al-Abdullah, *below,* as Salim al-Sabah, was serving as prime minister when Iraq invaded Kuwait in 1990. The ruling family returned to power after defeated Iraqi forces withdrew from Kuwait in February 1991.

the country from Iraq's grip.

The Persian Gulf War of 1991 caused immense human suffering and property damage in Kuwait. Hundreds of thousands of people were killed or wounded or became refugees.

In October 1992, Kuwait elected a new 50-member Parliament. In addition, the constitution was reinstated.

Kuwait is a small desert land at the tip of the Persian Gulf that sits atop huge oil reserves. Iraq and Saudi Arabia border Kuwait. Iran lies to the northeast. The area was the center of major conflict in the early 1990's.

Languages spoken:
Arabic (official)
English
Religions:
Muslim 85% (Sunni 45%, Shiah 40%)
Christian
Hindu
Parsi

TECHNOLOGY

Radios per 1,000 people: 624

Televisions per 1,000 people: 486

Computers per 1,000 people: 130.6

ECONOMY

Currency: Kuwaiti dinar

Gross national income (GNI) in 2000: $35.8 billion U.S.

Real annual growth rate (2000): 1.7%

GNI per capita (2000): $18,030 U.S.

Balance of payments (2000): $14,865 million U.S.

Goods exported: Oil and refined products, fertilizers

Goods imported: Food, construction materials, vehicles and parts, clothing

Trading partners: Japan, United States, India, United Kingdom

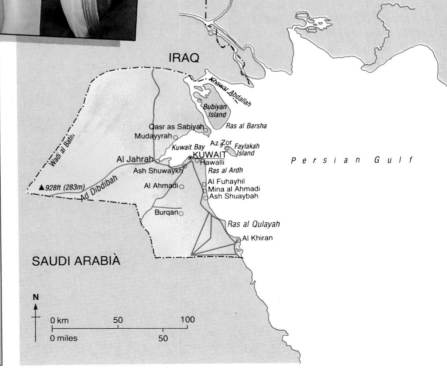

IRAQ

Khawr Abdallah

Bubiyan Island

Ras al Barsha

Qasr as Sabiyah
Mudayyrah

Az Zor
Faylakah Island

Wadi al Batin

Kuwait Bay

KUWAIT

Al Jahrah

Hawalli
Ras al Ardh

Ash Shuwaykh

Persian Gulf

▲928ft (283m)

Ad Dibdibah

Al Ahmadi

Al Fuhayhil
Mina al Ahmadi
Ash Shuaybah

Burqan

Ras al Qulayah

SAUDI ARABIA

Al Khiran

N

0 km   50   100
0 miles   50

Economy

Kuwait draws most of its income from a single source—oil. The government also receives large amounts of money from earnings on investments in the United States and other countries.

Kuwait has used most of its oil wealth to support its welfare system and modernize the country. The government has also given financial aid to other Arab countries through the Kuwait Fund for Arab Economic Development. The fund has also provided aid to non-Arab countries in Africa and Asia.

Although the oil industry produces vast wealth for Kuwait, it does not employ many people. Most of the work is done by machines and the majority of the jobs available are held by noncitizens because most Kuwaitis lack the necessary education and skills to perform them. Many Kuwaitis therefore depend on welfare to live. However, the government is spending money on education and job training so that more Kuwaitis will be able to handle skilled work in the future.

Formation and discovery of oil

Millions of years ago, much of what is now the Middle East lay under a sea, according to most geologists. Countless tiny life forms in that sea sank to the sea floor when they died. There they became trapped in *sediments* (particles of mud and sand). These sediments piled up, sank deeper and deeper under further layers of mud, and began to change chemically under the pressure and the heat. Gradually, they decayed into *hydrocarbons*—the basic ingredients of oil and natural gas.

Over time, the mixture of oil and gas seeped upward through *pores*, or natural passageways, in the rock. Meanwhile, the sea above drew back until much of the oil and gas eventually lay under desert land. There, the deposits lay undisturbed until the 1930's.

In 1934, Kuwait's ruler allowed the Kuwait Oil Company, a joint American-British group, to explore for oil. Drilling began in 1936 and showed that vast quantities of petroleum lay under the desert of Kuwait. When World War II ended in 1945, Kuwait became a major petroleum exporter and an extremely wealthy country. In 1975, the government *nationalized*, or took control of, the Kuwait Oil Company and now owns almost all of the industry in the country.

The politics of oil

As a member of the Organization of Petroleum Exporting Countries (OPEC), Kuwait has sometimes used its valuable oil to influence world affairs. For about two months in 1967, in response to the Six-Day War, Kuwait cut off oil shipments to the United States and other Western countries that supported Israel. In 1973, it joined other Arab nations in stopping shipments to the United States and the Netherlands because of another Arab-Israeli war.

Kuwait's valuable oil made the country a focal point of international attention during the late 1980's and early 1990's. In 1987, Kuwait's support of Iraq in its eight-year conflict with Iran triggered attacks by Iran on Kuwaiti oil tankers in the Persian Gulf.

In 1990, Iraqi president, Saddam Hussein, ordered his troops to invade Kuwait, in part to acquire its oil wealth. Iraq had suffered serious economic damage during the Iran-Iraq War, and, by seizing Kuwait, Hussein also sought to eliminate the huge debt Iraq owed Kuwait.

The Persian Gulf War, which was fought in early 1991 to liberate Kuwait from Iraq, created tremendous problems for Kuwait's oil industry. Iraqi troops badly damaged almost half of Kuwait's 1,300 oil wells. In most cases, they set them on fire, which produced a thick black smoke—and severe air pollution—throughout the region.

After the war, Kuwait began the huge task of repairing its oil storage, refining, and transportation facilities. In late 1991, two of Kuwait's three major refineries remained incapacitated. By early 1992, Kuwait had restored only half of its oil refining capacity.

After the war, Kuwait was exempt from OPEC production quotas so it could rebuild its economy. By early 1993, Kuwait was able to produce 2 million barrels of oil a day. In February of that year, Kuwait agreed to hold production down to 1.6 million barrels a day in order to raise the price of oil. Kuwaiti leaders initially resisted the quota, which was proposed by ministers from OPEC, maintaining that Kuwait should continue to be exempt from OPEC quotas.

A Kuwaiti buys gasoline for his car at a Kuwait National Petroleum Company (KNPC) service station. Formed in 1960, KNPC was the first non-foreign-owned oil company in Kuwait.

Pipelines, *left,* carry Kuwait's petroleum across the desert to the Persian Gulf. The natural gas that accompanies the petroleum is often recovered for export, but here it is being burned off.

Water-storage towers resembling minarets rise near the shore of Kuwait Bay. A large underground reservoir, discovered in 1960, supplements the water produced by distillation plants.

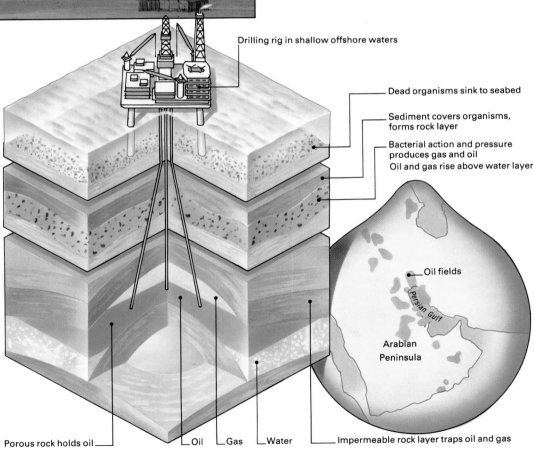

Drilling rig in shallow offshore waters

Dead organisms sink to seabed

Sediment covers organisms, forms rock layer

Bacterial action and pressure produces gas and oil

Oil and gas rise above water layer

Oil fields

Persian Gulf

Arabian Peninsula

The formation of oil in the Persian Gulf region began with the decay of tiny marine plants and animals many millions of years ago. Pressure applied by layers of sediments aided the oil formation.

Porous rock holds oil

Oil

Gas

Water

Impermeable rock layer traps oil and gas

Kyrgyzstan

Kyrgyzstan is a mountainous country bordered by China in the southeast, Tajikistan in the south, Uzbekistan in the southwest, and Kazakhstan in the north and northwest.

The Alay and Tian Shan mountains cover much of Kyrgyzstan. Nestled in the Tian Shan is Issyk-Kul Lake, one of the world's largest and deepest mountain lakes.

The history of Kyrgyzstan is part of the history of Turkestan, the historic region in central Asia to which it belongs. Around the time of Christ, much of this region belonged to the Chinese Empire. In the 500's and 600's, what is now Kyrgyzstan was a center of power for some Turkic tribes. Turkestan then underwent a period of Chinese rule. Beginning in the 700's, Arab invaders introduced Islam into the region.

The Tatars swept through the region in the 1200's and 1300's. It became part of the Russian Empire in the late 1800's. Kyrgyzstan was made an autonomous region within the Russian Republic of the Soviet Union in 1924, and it became a separate Soviet republic in 1936.

Kyrgyzstan remained under the strict control of the Soviet central government until the early 1990's. In the midst of political upheaval in the Soviet Union following an attempted coup in August 1991, Kyrgyzstan declared its independence.

In December 1991, Soviet President Mikhail Gorbachev and Russian President Boris Yeltsin agreed to dissolve the Soviet Union by the end of the year and replace it with a new association called the Common-

Kyrgyzstan is a land of towering, snow-capped peaks, alpine meadows, and deep valleys.

FACT BOX

COUNTRY

Official name: Kyrgyz Respublikasy (Kyrgyz Republic)
Capital: Bishkek
Terrain: Peaks of Tian Shan and associated valleys and basins encompass entire nation
Area: 76,641 sq. mi. (198,500 km²)

Climate: Dry continental to polar in high Tian Shan, subtropical in southwest (Fergana Valley), temperate in northern foothill zone
Main river: Naryn
Highest elevation: Jengish Chokusu (Peak Pobedy), 24,406 ft. (7,439 m)
Lowest elevation: Kara-Darya, 433 ft. (132 m)

GOVERNMENT

Form of government: Republic
Head of state: President
Head of government: Prime minister
Administrative areas: 6 oblastlar, 1 shaar (city)
Legislature: Zhogorku Kenesh (Supreme Council) consisting of the Assembly of People's Representatives with 70

members serving five-year terms and the Legislative Assembly with 35 members serving five-year terms
Court system: Supreme Court, Constitutional Court, Higher Court of Arbitration
Armed forces: 9,200 troops

PEOPLE

Estimated 2002 population: 4,784,000
Population growth: 1.43%
Population density: 62 persons per sq. mi. (24 per km²)
Population distribution: 66% rural, 34% urban
Life expectancy in years:
Male: 59
Female: 68
Doctors per 1,000 people: 3.0
Percentage of age-appropriate population enrolled in the following educational levels:
Primary: 104*
Secondary: 86
Further: 30

wealth of Independent States (CIS). The CIS had been established the previous week as a loose confederation of former Soviet republics. Kyrgyzstan agreed to join the CIS.

About 50 per cent of the people who live in Kyrgyzstan belong to the Kyrgyz nationality group. The Kyrgyz speak one of the Turkic languages and many are Muslims, especially in the south. Most Kyrgyz are short people with yellowish complexions and black hair. Ethnic tensions between the Kyrgyz and another group, the Uzbeks, led to violence in the 1990's. Hundreds of people died in the clashes.

For centuries, many Kyrgyz were nomads, wandering herders who raised their sheep in the mountain meadows in the summer and brought them down to the foothills in winter. However, during the 1930's, the Soviet government forced many of the Kyrgyz to live and work on collective farms. Despite the scarcity of rainfall in Kyrgyzstan, crops such as cotton and tobacco prosper, helped by the use of extensive irrigation systems.

Kyrgyzstan has large deposits of antimony, coal, lead, natural gas, and oil. Under Soviet rule, these natural resources became very important in transforming Kyrgyzstan's agrarian economy into one based on mining and the production of raw materials.

The rugged peaks and rivers of the Tian Shan, which cover much of Kyrgyzstan, are home to abundant fish and wildlife. This towering mountain system runs for nearly 1,500 miles (2,400 kilometers) and includes some of the world's largest glaciers. Tian Shan means *Heavenly Mountains.*

Languages spoken:
Kirghiz or Kyrgyz (official)
Russian (official)

Religions:
Muslim 75%
Russian Orthodox 20%

Enrollment ratios compare the number of students enrolled to the population which, by age, should be enrolled. A ratio higher than 100 indicates that students older or younger than the typical age range are also enrolled.

TECHNOLOGY

Radios per 1,000 people: 111

Televisions per 1,000 people: 49

Computers per 1,000 people: N/A

ECONOMY

Currency: Kyrgyzstani som

Gross national income (GNI) in 2000: $1.3 billion U.S.

Real annual growth rate (1999–2000): 5.0%

GNI per capita (2000): $270 U.S.

Balance of payments (2000): -$77 million U.S.

Goods exported: Cotton, wool, meat, tobacco, gold, mercury, uranium, hydropower, machinery, shoes

Goods imported: Oil and gas, machinery and equipment, foodstuffs

Trading partners: Germany, Russia, Kazakhstan, Uzbekistan

Industrial development led large numbers of European immigrants—especially Russians—to settle in northern Kyrgyzstan, where Bishkek, the capital and industrial center, is located.

Because the economy was focused on mining and the production of raw materials, the lack of a consumer goods industry has often caused shortages in the stores. The development of heavy industry has also resulted in environmental problems.

Although modern development has forever changed the lives of the Kyrgyz, some old traditions have endured. For example, the folklore of the Kyrgyz lives on in *Manas,* the longest oral chronicle of its kind in the world today. The tale describes the many adventures of the hero Manas the Strong, including how he defended his people against the threat of conquerors.

Laos

Laos, a tropical land of mountains and thick forests, is the only landlocked country on the Southeast Asian peninsula. It is bordered by Myanmar and China to the north, Cambodia to the south, Vietnam to the east, and Thailand to the west. The capital of Laos is Vientiane.

Laos lies in the Mekong Basin, between the Mekong River and the Annamite Range. The Mekong River, the chief means of transportation in Laos, also waters the country's fertile lowland.

The people of Laos belong to two language groups—the Sino-Tibetan from China and the Mon-Khmer from Southeast Asia. The Sino-Tibetan group includes the Lao, Hmong (also called Meo), and Tai peoples. The Mon-Khmer group includes the Kha peoples. Many of these ethnic groups have lived in Laos for more than 1,000 years.

Land of a million elephants

Ancestors of the Lao and Tai probably moved into Laos in the A.D. 800's from southwest China. They established a number of small states on the Mekong River, notably Muong Swa (now Louangphrabang). In 1353, the ruler of Muong Swa united most of what is now Laos to form the kingdom of *Lan Xang* (Land of a Million Elephants).

The kingdom remained unified for almost 350 years. About 1700, however, quarreling among powerful groups in Lan Xang destroyed the kingdom's unity, and it was divided into three principalities—Louangphrabang, Vientiane, and Champasak.

French Indochina and independence

By the late 1800's, the French had gained considerable power and influence throughout *Indochina*—that is, the eastern half of the Southeast Asian peninsula, which includes Laos, Cambodia, and Vietnam. By the 1880's, Cambodia and Vietnam had become French protectorates. By 1917, Louangphrabang was also a protec-

torate, and Champasak and Vientiane formed a colony in French Indochina.

In 1940, after France surrendered to Germany in World War II (1939-1945), the Japanese moved their forces into Indochina. They persuaded the rulers of Vietnam, Cambodia, and Laos to declare their countries independent under pro-Japanese governments. The government of Laos was headed by three princes—Phetsarat, Souvanna Phouma, and Souphanouvong.

In 1949, when the French declared Laos an independent kingdom, the three princes split into rival factions. Souphanouvong moved to northeastern Laos, and established the Communist-inspired Pathet Lao movement with North Vietnamese Communist leader Ho Chi Minh.

Civil war

In 1960, Captain Kong Le, a Laotian army officer, overthrew Laos' pro-Western government, and civil war broke out. A coalition government was set up in 1962 with Souvanna Phouma as prime minister, but Souphanouvong withdrew from the government in 1963 and fighting broke out again. By 1970, Souvanna Phouma's government troops controlled only western Laos, while Souphanouvong and the Pathet Lao held eastern Laos.

During the Vietnam War, North Vietnam used the Ho Chi Minh Trail in Laos and Cambodia to move troops and supplies into South Vietnam. United States planes bombed the trail and other areas in Laos. In 1971, South Vietnamese troops attacked Communist supply routes in Laos, but Communist forces drove them out.

The Laotian government and the Pathet Lao agreed to a cease-fire in 1973 and formed a coalition government, but by 1975 the government was dominated by Communists. South Vietnam and Cambodia fell to the Communists in April 1975, and later that year the Pathet Lao seized control of Laos, so that country also became a Communist state.

Laos Today

In 1975, Souphanouvong was installed as the president of the Lao People's Democratic Republic, but this was largely an honorary post. Real power lay with the country's Communist organization, the Lao People's Revolutionary Party (LPRP). The secretary-general of the LPRP became prime minister of Laos.

Socialism

Soon after coming to power, the LPRP attempted to restructure Laos according to socialist principles. The country's few industries were taken over by the government, and many small family-run farms were consolidated into state-run collective farms.

The government's task was hindered by the Laotian economy, which was largely undeveloped. Agriculture had traditionally been the country's chief economic activity, but old-fashioned equipment and methods held down agricultural output. In addition, many farmers preferred to destroy or abandon their farms rather than submit to *collectivization,* a government program in which the state owns the land and controls the workers.

Many Laotian people also resisted other repressive government policies, including seizures of private property and the suppression of religion. Thousands of people fled the country, most of them to Thailand.

A street in Vientiane, *center,* is bathed in the pink light of morning as many people cycle to the capital's markets. Private enterprise reemerged following the economic reforms of the 1980's, along with a greater availability of farm products.

Reforms

By 1980, the Laotian leaders had launched a number of economic reforms. The government halted the creation of new collective farms, and some private enterprise was allowed in business. In addition, farmers were paid more for agricultural goods. In 1981, rice production reached record levels. The Lao People's Democratic Republic made progress toward self-sufficiency in food production during the 1980's.

FACT BOX

LAOS

COUNTRY

Official name: Sathalanalat Paxathipatai Paxaxon Lao (Lao People's Democratic Republic)
Capital: Vientiane
Terrain: Mostly rugged mountains; some plains and plateaus
Area: 91,429 sq. mi. (236,800 km²)

Climate: Tropical monsoon; rainy season (May to November), dry season (December to April)
Main river: Mekong
Highest elevation: Phou Bia, 9,242 ft. (2,817 m)
Lowest elevation: Mekong River, 230 ft. (70 m)

GOVERNMENT

Form of government: Communist state
Head of state: President
Head of government: Prime minister
Administrative areas: 16 khoueng (provinces), 1 kampheng nakhon (municipality), 1 khethpiset (special zone)

Legislature: National Assembly with 99 members serving five-year terms
Court system: People's Supreme Court
Armed forces: 29,100 troops

PEOPLE

Estimated 2002 population: 5,709,000
Population growth: 2.5%
Population density: 62 persons per sq. mi. (24 per km²)
Population distribution: 83% rural, 17% urban
Life expectancy in years: Male: 51 Female: 55
Doctors per 1,000 people: 0.2
Percentage of age-appropriate population enrolled in the following educational levels: Primary: 111* Secondary: 33 Further: 3
Languages spoken: Lao (official) French

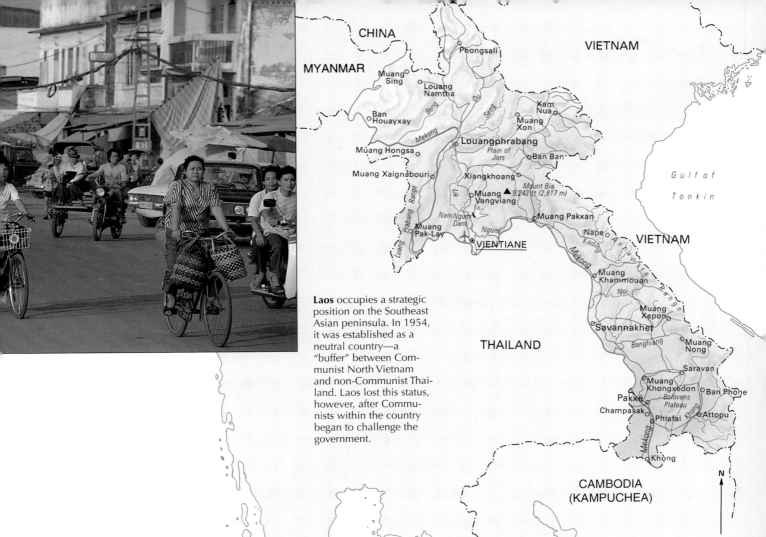

Laos occupies a strategic position on the Southeast Asian peninsula. In 1954, it was established as a neutral country—a "buffer" between Communist North Vietnam and non-Communist Thailand. Laos lost this status, however, after Communists within the country began to challenge the government.

English
various ethnic languages

Religions:
Buddhist 60%
Animist

Enrollment ratios compare the number of students enrolled to the population which, by age, should be enrolled. A ratio higher than 100 indicates that students older or younger than the typical age range are also enrolled.

TECHNOLOGY

Radios per 1,000 people:
148
Televisions per 1,000 people: 10
Computers per 1,000 people: 2.6

ECONOMY

Currency: New kip
Gross national income (GNI) in 2000:
$1.5 billion U.S.
Real annual growth rate (1999–2000):
5.7%
GNI per capita (2000): $290 U.S.
Balance of payments (2000):
$90 million U.S.
Goods exported: Wood products, garments, electricity, coffee, tin
Goods imported: Machinery and equipment, vehicles, fuel
Trading partners: Thailand, Vietnam, Japan, Germany

In 1989, LPRP leaders took steps to improve Laos's economic and diplomatic relations with the rest of the world and sought foreign trade and investment.

In March 1989, the first national elections were held in Laos since the Communists came to power in 1975. Voters had a choice of 121 candidates competing for 99 seats in the country's legislative body, the Supreme People's Assembly. However, with no opposition parties allowed, most of the candidates came from the LPRP, and 65 party members were elected.

President Kaysone Phomvihane, Laos's ruler since the Communist takeover in 1975, died in late 1992. Nouhak Phoumsavan, a Communist hardliner, was elected president by a special session of the Assembly.

In 1994, Laos opened its first permanent bridge across the Mekong River, which runs along its border with Thailand.

Land and Economy

Densely forested mountains and rugged plateaus cover much of Laos, rising from 500 to 4,000 feet (150 to 1,200 meters) in the north and along the eastern border. South of the Plain of Jars—a wide plateau of rolling hills and wooded areas—the country's highest peak, Mount Bia, soars to a height of 9,242 feet (2,817 meters).

Southern Laos, narrower and less mountainous than the northern part of the country, is bordered on the east by the Annamite Range, which separates Laos from Vietnam and slopes gently westward to the Mekong Basin. The Bolovens Plateau lies in the extreme southern region of Laos.

The most productive farmland in Laos lies along the Mekong River and its tributaries, and most of the country's rice is grown in this fertile region. The river and its tributaries are also the chief means of transportation in Laos.

Laos has a tropical climate marked by humidity, high temperatures, and seasonally heavy rainfall. From May to September, southwest monsoons bring up to 10 inches (25 centimeters) of rain a month, and temperatures average about 82° F. (28° C). From November to March, when rainfall averages less than 1 inch (2.5 centimeters) a month, temperatures average about 70° F. (21° C).

The warm, wet climate of Laos makes the country an ideal place for agriculture. But only about 4 per cent of Laos' total land area is cultivated annually. In addition, old-fashioned equipment and methods keep farm production low.

The land is also rich in mineral resources. However, these have not been developed. The country also lacks manufacturing industries. All these factors combine to make Laos one of the poorest countries in the world.

Agriculture is the chief economic activity in Laos, and about 80 per cent of the people work on the land. Rice, which is grown mainly in the fertile river valleys, is the country's main crop. Laos produces enough rice for export in years of plentiful rain. Farmers also grow coffee, corn, cotton, and tobacco, and raise livestock. In the mountains along the Myanmar and Thailand borders, some farmers grow opium for export. The Laotian government, however, has taken steps to stamp out this illegal crop.

Laos has rich deposits of gold, gypsum, lead, silver, tin, and zinc, but like the country's industrial development, mining is hampered by a lack of electrical power and inadequate transportation. Most roads are impassable during the rainy season, and Laos has no railroads. However, a rail line links Nong Khai—across the Mekong River from Vientiane—with Bangkok, Thailand. In many areas, airplanes are the only means of moving supplies.

The country is also rich in teak and other valuable woods—forestry products make up about 30 per cent of Laos' total exports. However, much of the country's forestland is being cleared for farming.

Laos depends on foreign aid to develop its economy. Some of these funds have been used for forestry, irrigation, road and bridge construction, and hydroelectric power projects.

A square in Louangphrabang, once the royal capital of Laos, is shaded by lush vegetation, *right.* The city lies in the fertile flood plain along the Mekong River, northwest of the high plateau known as the Plain of Jars.

Overflowing with passengers and luggage, a bus prepares to pull out of Vientiane, *far right.* Transportation in Laos is poor, with most roads open only during the dry season.

Dense forest lines a stream in the Mekong Basin, *right,* where much of the nation's rice is grown. Laos usually produces at least enough rice to feed its own people.

Paddy rice is grown all over Asia in much the same way. In the first stage, farmers plant rice seeds, which have been saved from the previous season's crop, in a small seedbed **(1).** While the seeds are sprouting, farmers plow the main paddy field **(2).** The young rice plants are transplanted into the field after it has been flooded **(3).** Farmers weed the field while the rice grows, and ducks are often brought in to eat insects that damage the plants **(4).** When the rice ripens, the water is drained off by a simple irrigation system **(5).** A deep golden color indicates that the rice is ready for harvesting **(6).** After the harvest, the rice is tied in bundles and dried in the sun before being taken to the *threshing area,* where the grain is separated from the rest of the plant **(7).**

Workers in a tannery use a sewing machine to stitch animal hides. The few manufacturing industries in Laos depend mainly on livestock and other farm products for their raw materials.

People

Most of the people of Laos live in the fertile lowlands that stretch along the Mekong River. However, the nation is sparsely populated, with an average of only about 46 people per square mile (18 per square kilometer).

The people of Laos belong to four major ethnic groups—the Lao, Tai, Hmong (or Meo), and Kha. The Lao, Tai, and Hmong peoples belong to the Sino-Tibetan language group of China, while the Kha, the original inhabitants of Laos, belong to the Mon-Khmer language group of southeast Asia.

The Lao, who have given their name to the country and to its official language, make up about 50 per cent of the population and serve as the country's political and social leaders. The Kha have traditionally been treated as little more than slaves by the Lao, but the Communist-led Pathet Lao have worked to improve the social position of the Kha since they took over Laos in 1975.

People of the lowlands

Most Laotians are rice farmers who live in villages along the Mekong River and follow a traditional way of life. The villagers live in bamboo houses with thatched roofs. The houses are built on stilts to protect them from flooding during the rainy season.

Most Laotians are Buddhists, and village life centers around the *pagoda* (temple), where a Buddhist monk conducts religious ceremonies. The pagoda may also be used as a schoolhouse, or as a meeting place where the village headman leads a discussion on local affairs. The social life of the villagers centers around Buddhist festivals and holidays.

About 19 per cent of the Laotians live in urban areas and work in trade or industry. Many Westernized Lao and most of the foreigners in Laos live in the capital city of Vientiane, which has a population of 264,277 and ranks as Laos' largest city as well as an important trading center.

The mountain peoples

Among the Tai peoples, who live mainly in the mountain valleys of northern Laos, there are a number of different tribes—

The Hmong peoples, *below,* live in northern Laos, near the border of Myanmar and Thailand. Silver coins on a Hmong headdress indicate wealth.

Buddhist monks, *bottom,* relax in a rural monastery. When the Pathet Lao came to power, they discouraged religious practice, but the deep resentment of the Laotians caused many restrictions against religion to be lifted.

Lao villagers wade across the shallow Ou River in central Laos near Louangphrabang. The river waters the villagers' crops and provides a bountiful catch of fish.

Laotian women prepare food offerings for the monks at a Buddhist pagoda. Lao Buddhists believe that they will earn merit in a future life by maintaining the pagoda and feeding the monks.

each with its own dialect. Some of the tribes are named after the predominant color of the clothing woven and worn by the women. Often, several Tai villages are linked together in a unit, led by a chieftain, known as a *muong* (nation).

By contrast, the Hmong peoples, who live on the higher mountain slopes, have no unit of social organization larger than the village, which is led by a headman. The Hmong clear small forest areas to raise livestock and grow crops. One of their main crops is opium—an illegal but highly profitable product that is exported all over Southeast Asia.

Hmong are fiercely independent people who have little in common—either socially or politically—with the other peoples of Laos, or even with people of other Hmong villages. During the civil war years of the 1960's and 1970's, some Hmong people supported the Communist-led Pathet Lao while others waged a guerrilla campaign against the Communists. The Hmong who resisted Communism suffered severe persecution after the Pathet Lao came to power in 1975, and those who

supported the U.S. presence continue to be oppressed.

Religion and education

Most Laotians follow the Theravada school of Buddhism. In Laos, as in other Southeast Asian countries, many Buddhists combine their religious practice with a belief in spirits.

Although the Pathet Lao discouraged the practice of religion when they first came to power, the government today is more tolerant of religion. In an effort to align Communist principles and Buddhist beliefs, the Communists portrayed Buddha as a "revolutionary" who, by leaving his home, questioned the value of wealth and possessions.

In some areas, the monks in the village pagoda provide the only education for the local children because many villages have no school. Under the Pathet Lao, education is state-controlled, but money for educational facilities is still limited. In the 1980's, only about 56 per cent of Laotian children between the ages of 6 and 17 were attending school.

Latvia

A Soviet republic from 1940 until it gained its independence in 1991, Latvia lies in northern Europe along the Baltic Sea. A large plain covers most of Latvia, and the landscape consists chiefly of low hills and valleys. Many small lakes and swamps dot the countryside, and forests cover about 40 per cent of the land.

History

About the time of Christ, the original inhabitants of Latvia were forced out by invaders who became the ancestors of present-day Latvians. The area later came under attack from the Vikings in the 800's and the Russians in the 900's. The Teutonic Knights took control in the 1200's and ruled Latvia as part of a larger state called Livonia.

By 1562, most of Latvia fell under Polish and Lithuanian control, while the rest was a German-ruled duchy. Sweden conquered northern Latvia in 1621, but by 1800, the Russians had gained control of the country. In 1918, Latvia declared its independence, and a new constitution, adopted in 1922, established a democratic government.

However, Latvian democracy did not survive the worldwide depression of the 1930's.

In 1936, the nation's president seized power and reduced the role of parliament and the rights of the country's political parties. In 1940, the Soviet Union forced Latvia to sign a treaty under which the Soviets built military bases in the country. In June 1940, Soviet troops occupied Latvia, and Latvian Communists took over the government.

Through the years, the Latvians expressed their opposition to Soviet rule. A strong nationalist movement took shape during the late 1980's, when Latvian reformers established the People's Front, a large non-Communist organization. On May 4, 1990, Latvia's parliament declared the restoration of Latvian independence and called for a gradual separation from the Soviet Union. However, the Soviet central government declared the action illegal.

In August 1991, conservative Communist leaders in the Soviet Union attempted to overthrow the central government. However, the coup failed. During the political upheaval that followed, Latvia declared its independence, and in September of 1991, the Soviet government recognized Latvia's independence.

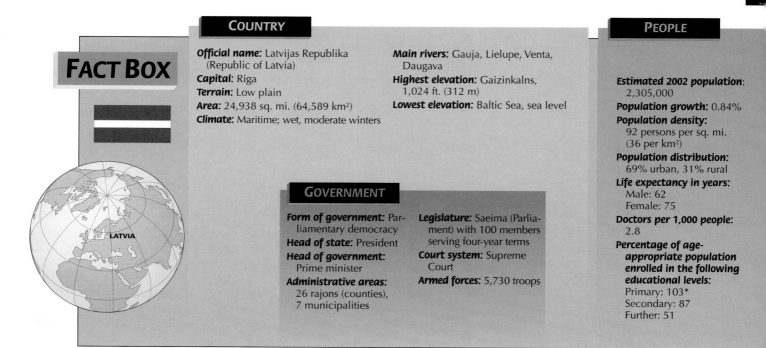

FACT BOX

COUNTRY

Official name: Latvijas Republika (Republic of Latvia)
Capital: Riga
Terrain: Low plain
Area: 24,938 sq. mi. (64,589 km²)
Climate: Maritime; wet, moderate winters

Main rivers: Gauja, Lielupe, Venta, Daugava
Highest elevation: Gaizinkalns, 1,024 ft. (312 m)
Lowest elevation: Baltic Sea, sea level

GOVERNMENT

Form of government: Parliamentary democracy
Head of state: President
Head of government: Prime minister
Administrative areas: 26 rajons (counties), 7 municipalities

Legislature: Saeima (Parliament) with 100 members serving four-year terms
Court system: Supreme Court
Armed forces: 5,730 troops

PEOPLE

Estimated 2002 population: 2,305,000
Population growth: 0.84%
Population density: 92 persons per sq. mi. (36 per km²)
Population distribution: 69% urban, 31% rural
Life expectancy in years: Male: 62 Female: 75
Doctors per 1,000 people: 2.8
Percentage of age-appropriate population enrolled in the following educational levels: Primary: 103* Secondary: 87 Further: 51

People

About 52 per cent of the people who live in Latvia are Latvians, a separate nationality group which has its own culture and language. The Latvian language is one of the oldest in Europe. It is related to Sanskrit, a language of ancient India. Latvia has about 900,000 ethnic Russians. In June 1994, Latvia passed a law that made it very difficult for people who are not ethnic Latvians to become citizens. The law drew protests from Russia and other European countries. Latvia eased the law in response.

A child gazes in wonder at an oil well on a wooded Latvian plain, *above left.* Today, industrial waste poses an enormous threat to the natural environment of Latvia.

A graceful suspension bridge spans the Daugava in Riga, the capital and largest city of Latvia. Riga is a major shipping center and accounts for more than half of Latvia's industrial output.

Latvia, *map below,* lies in northern Europe on the eastern shore of the Baltic Sea. Along with Estonia and Lithuania, it is one of the three Baltic States.

Languages spoken:
Latvian (official)
Lithuanian
Russian

Religions: Lutheran
Roman Catholic
Russian Orthodox

Enrollment ratios compare the number of students enrolled to the population which, by age, should be enrolled. A ratio higher than 100 indicates that students older or younger than the typical age range are also enrolled.

TECHNOLOGY

Radios per 1,000 people:
695

Televisions per 1,000 people: 789

Computers per 1,000 people: 140.3

ECONOMY

Currency: Latvian lat

Gross national income (GNI) in 2000: $6.9 billion U.S.

Real annual growth rate (1999–2000): 6.6%

GNI per capita (2000): $2,920 U.S.

Balance of payments (2000): -$494 million U.S.

Goods exported: Wood and wood products, machinery and equipment, metals, textiles, foodstuffs

Goods imported: Machinery and equipment, chemicals, fuels

Trading partners: Germany, Russia, United Kingdom, Finland

Lebanon

Located between the Mediterranean world to the west and Asia to the east, Lebanon is a Middle East land that has developed in its own unique way. The sea has helped make Lebanon an important trading region for thousands of years. As a center for trade, it was exposed to people of many cultures. Some left their mark on Lebanon; others invaded and even ruled the country.

However, the rugged mountains that lie within the country have helped protect the people of the region and have enabled Lebanon to survive with an identity of its own. Its land and location help explain why Lebanon today has an Arab culture influenced heavily by Western and Christian ideas.

Ancient seafarers and Roman rule

Lebanon's history of sailing and trading can be traced back 5,000 years. About 3000 B.C., the Phoenicians moved into the region and established powerful city-states along the coast. The Phoenicians were skillful sailors who traded with ancient Egypt and other regions and founded colonies along the shores of the Mediterranean Sea. Evidence shows that the Phoenicians sailed around Africa thousands of years before the Portuguese did.

Beginning about 1800 B.C., other foreign powers began to conquer and control the city-states of Lebanon. Egyptians, Hittites, Assyrians, Babylonians, and Persians all included the region as part of their ancient empires. Alexander the Great of Macedonia conquered Lebanon in 332 B.C.

Lebanon became part of the vast Roman Empire in 64 B.C. Roman ruins, such as the temples at Baalbek in the Bekaa Valley, still stand in the countryside. The Romans eventually adopted Christianity, and so did many Lebanese.

Christians and Muslims

In the A.D. 600's, Arab Muslims invaded the region. As they did in other nations in North Africa and the Middle East, the Arabs converted many people in Lebanon to the religion of Islam. However, Islam replaced

Christianity only along Lebanon's coast. Most of the Lebanese who lived in the mountains remained Christians.

Christian crusaders from Europe invaded Lebanon about 1100, hoping to win back control of the Holy Land south of Lebanon from the Muslims. The crusaders established friendly relations with the Christians in the mountains, and the Lebanese Christians became open to European influences and were more willing to be approached by Westerners. But Mameluke Muslims from Egypt drove the crusaders out of Lebanon about 200 years later.

Another powerful empire—that of the Ottoman Turks—conquered Lebanon in 1516 and ruled the region for more than 400 years. During part of this time, local rulers in Mount Lebanon—the largely Christian region in the central mountainous part of the country—were allowed a greater degree of self-rule.

When the Ottoman Empire was defeated in World War I (1914–1918), Great Britain and France occupied Lebanon. France was given control of the region's political affairs in 1922. The French united the Christians in Mount Lebanon and the Muslims along the coast under one government. Lebanon became a fully independent nation in 1943.

In the new nation, Christian and Muslim leaders agreed to share power. Lebanon kept its strong ties with Western nations and enjoyed increasing prosperity as a center of trade and finance. Lebanon also became an important financial center. About 100 banks—including branches of many foreign banks—operated within the country.

The differences between Christians and Muslims eventually led to trouble, however. In 1958, some Lebanese, mostly Muslims, rebelled against the government and its close alliance with the West. War flared up in the 1970's, and various factions fought off and on through the 1980's.

In June 1982, Israel invaded Lebanon and drove Palestine Liberation Organization (PLO) forces out of southern Lebanon. In 1985, Israel withdrew from all of Lebanon except a self-proclaimed security zone along its border.

A peace plan led most of Lebanon's private military groups, called *militias,* to disarm and disband in 1991. Although some violence continued, national reconstruction began.

Lebanon Today

The fighting that occurred in Lebanon until the early 1990's affected every aspect of Lebanese life. Thousands of people were killed, sections of cities were destroyed, and the economy was severely damaged.

According to its Constitution, Lebanon is a republic. The people elect members of the National Assembly to make the country's laws, and the Assembly members elect a president. The president then appoints a prime minister, who chooses a Council of Ministers to help run the government.

Traditionally, the president and prime minister have worked as a team. To maintain a power balance between the country's Muslims and Christians, the government has been made up of both groups in proportion. Also, a Christian has served as president, while a Sunni Muslim has been appointed prime minister.

This careful balance was disrupted when the Muslim population grew larger than the Christian population, and Muslims began demanding more power. Another disruption was the presence of the Palestine Liberation Organization (PLO) in the country. The PLO established bases in southern Lebanon to launch attacks on Israel to the south. The Lebanese Muslims supported the PLO, while the Christians opposed them.

FACT BOX

COUNTRY

Official name: Al Jumhuriyah al Lubnaniyah (Lebanese Republic)
Capital: Beirut
Terrain: Narrow coastal plain; Al Biqa' (Bekaa Valley) separates Lebanon and Anti-Lebanon Mountains
Area: 4,015 sq. mi. (10,400 km²)

Climate: Mediterranean; mild to cool, wet winters with hot, dry summers; Lebanon mountains experience heavy winter snows
Main rivers: Litani, Nahr Ibrahim, Orontes
Highest elevation: Qurnat as Sawda', 10,131 ft. (3,088 m)
Lowest elevation: Mediterranean Sea, sea level

GOVERNMENT

Form of government: Republic
Head of state: President
Head of government: Prime minister
Administrative areas: 5 mohafazat (governorates)
Legislature: Majlis Alnuwab, in Arabic, or Assemblee Nationale, in French, (both meaning National Assembly) with 128 members serving four-year terms

Court system: Courts of Cassation, Constitutional Council, Supreme Council
Armed forces: 67,900 troops

PEOPLE

Estimated 2002 population: 3,373,000
Population growth: 1.38%
Population density: 840 persons per sq. mi. (324 per km²)
Population distribution: 88% urban, 12% rural
Life expectancy in years:
Male: 69
Female: 74
Doctors per 1,000 people: 2.1
Percentage of age-appropriate population enrolled in the following educational levels:
Primary: 110*
Secondary: 89
Further: 38

A refugee camp in the suburbs of Beirut was patrolled by French soldiers as part of an international peacekeeping effort in Lebanon in 1982.

Lebanon is a small country at the eastern end of the Mediterranean Sea and the western end of Asia. The rugged Lebanon Mountains separate the country's sandy beaches and coastal plain from the fertile Bekaa Valley. The Anti-Lebanon Mountains lie east of the valley, along the Syrian border. The majestic cedars of Lebanon once covered the slopes of the mountains, but most have been cut down.

The civil war lasted from 1975 to 1976. In the spring of 1976, Syria sent troops into the country to restore order. Full-scale fighting ended later that year, but unrest continued.

Since 1976, Lebanon's Christians have fought among themselves as well as against the Muslims and the PLO. Muslim groups, too, have fought among themselves. The United Nations sent peacekeeping troops into Lebanon in 1978.

In June 1982, Israel invaded southern Lebanon, drove the PLO forces out, and seized western Beirut. Hundreds of civilian Palestinians in refugee camps were killed by Lebanese Christians. The United States, France, Italy, and the United Kingdom sent peacekeeping troops.

Christian President Amin Gemayel's term of office ended in September 1988, and the

Lebanese parliament failed to elect a successor. Gemayel appointed General Michel Aoun, the Christian commander of the Lebanese army, as prime minister of an interim military government. But since Lebanon already had a Muslim prime minister, Salim al-Huss, Muslims refused to recognize the appointment. As a result, two governments claimed to rule Lebanon.

Large-scale fighting broke out in 1989 between Aoun's forces and Syrian forces in Beirut. Fighting continued off and on until October 1990, when Aoun was ousted and his troops were disbanded. A Muslim, Omar Karami, was appointed as the new prime minister. By December, all opposing groups had withdrawn their troops from Beirut. Under a peace plan, most of Lebanon's factions disarmed in 1991.

In 1996, Israel launched an attack on southern Lebanon. Israel said it was retaliating against attacks on Israel by the Islamic extremist group Hezbollah. The U.S. brokered a peace agreement later that year. By then, hundreds of Lebanese had been killed or injured.

In 1998, Lebanon held its first presidential election since the end of its civil war.

Languages spoken:
Arabic (official)
French
English
Armenian

Religions:
Muslim 70%
Christian 30%
Jewish

Enrollment ratios compare the number of students enrolled to the population which, by age, should be enrolled. A ratio higher than 100 indicates that students older or younger than the typical age range are also enrolled.

TECHNOLOGY

Radios per 1,000 people:
687

Televisions per 1,000 people: 335

Computers per 1,000 people: 50.1

ECONOMY

Currency: Lebanese pound

Gross national income (GNI) in 2000: $17.4 billion U.S.

Real annual growth rate (1999–2000): 0.0%

GNI per capita (2000): $4,010 U.S.

Balance of payments (2000): -$3,065 million U.S.

Goods exported: Foodstuffs and tobacco, textiles, chemicals, metal and metal products, electrical equipment and products, jewelry, paper and paper products

Goods imported: Foodstuffs, machinery and transport equipment, consumer goods, chemicals, textiles, metals, fuels, agricultural foods

Trading partners: France, United States, Saudi Arabia, Italy, United Arab Emirates

People

Lebanon is home to more than 3 million people. About 90 per cent of the Lebanese people are Arabs.

This Arab population includes more than 400,000 Palestinian Arabs, refugees from the Arab-Israeli wars that have troubled the Middle East since 1948. These refugees had lived on land that is now part of Israel. Other ethnic minorities in Lebanon include Armenians, Assyrians, and Kurds.

Almost all Lebanese speak Arabic, the official language. Many also speak French or English. Armenian Lebanese speak the Armenian language.

About 60 per cent of the Lebanese are Muslims; almost all the rest are Christians. But a number of Muslim and Christian *sects* exist. A sect is a small religious group separated from an established church.

Most Lebanese Muslims belong to one of two major sects—the Sunni or the Shiah. Two smaller Muslim sects are the Ismailis and the Alawis. Druses practice a religion related to Islam. Some of the religious beliefs of the Ismailis, Alawis, and Druses are kept secret.

The majority of Lebanese Christians are members of the Maronite Church, an eastern branch of the Roman Catholic Church. Other Lebanese Christian groups include Greek Orthodox, Armenian Apostolics, and Jacobites.

Urban and rural family life

About 80 per cent of Lebanon's people live in urban areas. Most upper-class and middle-class Lebanese live in cities, and most are either Christians or Sunni Muslims.

Most poor Lebanese live in rural areas or in run-down city slums, and most are Shiah Muslims or Palestinian Arabs. Many of the Palestinian refugees live in crowded camps.

The family is important in all areas of Lebanese life. Family loyalty is highly valued, and well-to-do family members are expected to share their wealth with less fortunate relatives. Many wealthy city people help support family members in rural villages.

In business, employers prefer to hire relatives, and many businesses are family run.

Traditional clothing, like this woman's hat and scarf, is rare today in Lebanon, but some rural people still dress as their ancestors did. Most Lebanese wear Western-style clothes.

Shiite demonstrators, *right,* demand a greater voice in the affairs of their country. Most Shiah Muslims—the poorest Lebanese group—are farmers in rural southern Lebanon and the Bekaa Valley.

Lebanese Druses, like these women dressed in traditional mourning clothes, are a close-knit community. The Druse religion is an offshoot of Islam, with secrets that only certain Druses know.

Many religions and cultures mix in Lebanon. Representation in the government for each group was based on the 1932 census, which gave Christians an edge. Since then, the Muslim population has grown to become the majority.

Carrying on a long artistic tradition, a young Lebanese engraver, *above,* applies the final touches to a metal shield.

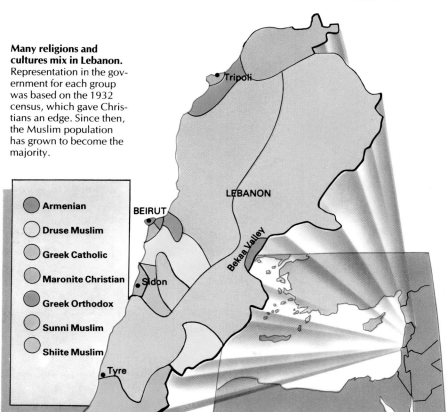

- Armenian
- Druse Muslim
- Greek Catholic
- Maronite Christian
- Greek Orthodox
- Sunni Muslim
- Shiite Muslim

Tripoli

BEIRUT

LEBANON

Sidon

Bekaa Valley

Tyre

Marriages are often arranged between families to help preserve the wealth and status of the families involved. In politics, families compete with each other for power.

The wealth and status of a family can also affect the kind of education a child receives. Lebanese law does not require children to go to school, but most parents send their children to both elementary and secondary school. More than half go to private schools, for which they must pay tuition, and the rest attend free public schools. Unfortunately, overall school attendance was disrupted by the fighting that plagued the country from the 1970's until the early 1990's.

Way of life

The Lebanese combine Western ways with their traditional or Arabic way of living. Many Lebanese houses have thick limestone walls and roofs made of orange tiles or earth, but in the cities, this type of house is rapidly being replaced by Western-style concrete houses and high-rise apartments.

Most Lebanese wear Western-style clothes, but some people in rural areas wear traditional Lebanese clothing. Some peasant women, for example, wear colorful, long dresses over ankle-length trousers. Some older Druse men wear jackets woven of many colors, and white headdresses.

The Lebanese enjoy such dishes as *dolmeh*—grape leaves stuffed with rice and ground lamb and spiced with cinnamon—served with *khobez* (flat, round wheat bread); or *hummus,* a tasty concoction of mashed chickpeas, *tahini* (sesame paste), garlic, and olive oil. Soft drinks are popular, as is Arabic coffee and a strong liquor called *arak.*

Many Lebanese people enjoy both Western and Arabic literature, music, and art. Lebanese artists are known for their beautiful silverware, brassware, jewelry, needlework, and colored glassware.

817

Lesotho

Lesotho is sometimes called the "Switzerland of southern Africa" because of its beautiful mountain scenery. The Drakensberg and the Maloti mountains cover most of this tiny nation. The only plains lie in the west, where about two-thirds of the people live.

The people and their work

But lovely Lesotho is also a poor country, with few industries and jobs. The most important economic activity in Lesotho is raising cattle, goats, and sheep. Farmers grow asparagus, beans, corn, peas, sorghum, and wheat, but overgrazing and overcultivation have damaged the soil.

The few Lesotho industries produce clothing, furniture, processed food, and textiles. Lesotho has diamond deposits, but diamond mining was halted in the 1980's when prices for the gems fell.

Most of Lesotho's people are black Africans called Basotho. Because Lesotho does not have enough jobs, many Basotho people go to South Africa to work. At any one time, about half of all the men are working in South African mines, factories, farms, or households. Generally, they work on contract for several months to two years, and the law requires them to deposit 60 per cent of their wages in the Lesotho Bank. The worker's family is allowed to withdraw half of the deposited funds to live on, but the other half must stay in the bank until the worker returns from South Africa. This system enables the government, which owns the Lesotho Bank, to invest funds in development projects.

Most Basotho people live in small rural villages, where family groups build their huts around a cattle *kraal,* or pen. The huts usually have mud or sod walls and thatched roofs, but wealthy Basotho may live in stone houses with roofs of tin or tile. Each village has a *khotla,* or meeting place, where men discuss village business.

The Basotho raise their crops on the land surrounding the village. All land is owned in common by the people and assigned by local chiefs.

History

The Basotho trace their history as a people to the late 1700's and early 1800's when tribal wars swept over southern Africa and many groups were almost completely wiped out. Some of the victims of this fighting fled into the highlands of what is now Lesotho, where they were given protection by an African chief named Moshoeshoe. By 1824, Moshoeshoe had about 21,000 followers, and he united them into the Basotho nation.

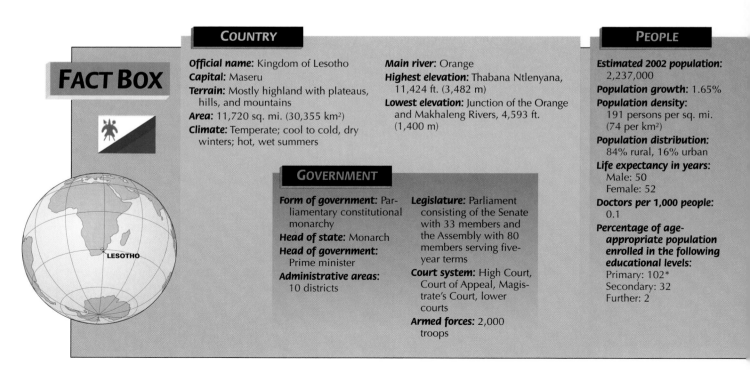

FACT BOX

COUNTRY

Official name: Kingdom of Lesotho
Capital: Maseru
Terrain: Mostly highland with plateaus, hills, and mountains
Area: 11,720 sq. mi. (30,355 km²)
Climate: Temperate; cool to cold, dry winters; hot, wet summers

Main river: Orange
Highest elevation: Thabana Ntlenyana, 11,424 ft. (3,482 m)
Lowest elevation: Junction of the Orange and Makhaleng Rivers, 4,593 ft. (1,400 m)

GOVERNMENT

Form of government: Parliamentary constitutional monarchy
Head of state: Monarch
Head of government: Prime minister
Administrative areas: 10 districts

Legislature: Parliament consisting of the Senate with 33 members and the Assembly with 80 members serving five-year terms
Court system: High Court, Court of Appeal, Magistrate's Court, lower courts
Armed forces: 2,000 troops

PEOPLE

Estimated 2002 population: 2,237,000
Population growth: 1.65%
Population density: 191 persons per sq. mi. (74 per km²)
Population distribution: 84% rural, 16% urban
Life expectancy in years: Male: 50 Female: 52
Doctors per 1,000 people: 0.1
Percentage of age-appropriate population enrolled in the following educational levels: Primary: 102* Secondary: 32 Further: 2

From 1856 to 1868, the Basotho were at war with South African settlers of Dutch descent called *Boers*. In 1868, the United Kingdom established the protectorate of Basutoland.

In 1966, Basutoland became the independent kingdom of Lesotho. The Basutoland National Party (BNP) leader Chief Leabua Jonathan became prime minister. Paramount Chief Motlotlehi Moshoeshoe II, great-grandson of Moshoeshoe, became king.

In January 1986, Jonathan was overthrown by a group of army officers, who in 1990 forced King Moshoeshoe II to leave office. They installed his oldest son as King Letsie III. King Letsie III dissolved the government and swore in a provisional council in 1994. When other nations denounced the move, the king agreed to restore the government and resign in favor of his father, the former King Moshoeshoe II. Moshoeshoe II died in 1996, and his oldest son returned to the throne as King Letsie III.

In elections held in 1998, a new party, the Lesotho Congress for Democracy (LCD), won 79 of the 80 seats in the Assembly. Opposition parties claimed that the elections were fixed, and protests broke out against the government. After much bloody fighting, an agreement was reached between the government and the opposition calling for new elections to be held within 18 months.

A Basotho wedding procession is traditionally accompanied by riders on horseback carrying flags. Basotho weddings are festive occasions, with much feasting, music, and dancing.

A typical Basotho boy begins work as a livestock herder at the age of five or six. In the east, boys may be away from home for months, searching for pasture with their herds. Because of their nomadic life, boys make up only about one-third of Lesotho's schoolchildren.

Languages spoken:
English (official)
Sesotho
Zulu
Xhosa
Religions:
Christian 80%
indigenous beliefs 20%

Enrollment ratios compare the number of students enrolled to the population which, by age, should be enrolled. A ratio higher than 100 indicates that students older or younger than the typical age range are also enrolled.

TECHNOLOGY

Radios per 1,000 people: 53

Televisions per 1,000 people: 16

Computers per 1,000 people: N/A

ECONOMY

Currency: Loti
Gross national income (GNI) in 2000: $1.2 billion U.S.
Real annual growth rate (1999–2000): 3.8%
GNI per capita (2000): $580 U.S.
Balance of payments (2000): -$151 million U.S.
Goods exported:
Mostly: manufactures (clothing, footwear, road vehicles)
Also: wool and mohair, food and live animals
Goods imported: Food; building materials, vehicles, machinery, medicines, petroleum products
Trading partners: South Africa Customs Union, North America, Asia

Lesotho, a mountainous country completely surrounded by South Africa, lies about 200 miles (320 kilometers) inland from the Indian Ocean. It was once called Basutoland, and ruled by the United Kingdom.

819

The Lesser Antilles

The West Indies are an island chain dividing the Caribbean Sea from the rest of the Atlantic Ocean. Stretching about 2,000 miles (3,200 kilometers), the chain forms a broad curve from near southern Florida to Venezuela's northern coast.

Three island groups make up the West Indies: the Bahamas in the north; the Greater Antilles, including Cuba, Jamaica, Hispaniola, and Puerto Rico; and the Lesser Antilles. The Lesser Antilles are a chain of small islands extending southeast of Puerto Rico from the Virgin Islands to Aruba.

The islands of the Lesser Antilles are noted for their great natural beauty. Sandy beaches and graceful palm trees line many of the coasts. The northern islands form a double chain—the western inner arc is made up of rugged, volcanic islands, while the outer eastern arc consists of low, flat-topped islands with limestone bases. The inner and outer arcs come together at the butterfly-shaped islands of Guadeloupe. The southern islands, from Trinidad and Tobago to Aruba, are made up of sedimentary rock.

The Windwards and the Leewards

Geographers divide the Lesser Antilles into the Windward Islands, which are exposed to the moist, northeast trade winds, and the Leeward Islands, which include the more sheltered downwind northerly and westerly islands.

The Windward Islands include Martinique, St. Lucia, Grenada, St. Vincent, and the Grenadine chain. About 15 main islands—including the Virgin Islands, Antigua and Barbuda, Dominica, St. Christopher and Nevis, and Guadeloupe, as well as many islets—make up the Leeward Islands.

Early history

Ciboney Indians were the earliest known prehistoric inhabitants of the West Indies.

Fishing crews prepare their nets before setting sail from a sheltered bay on St. Lucia. About 90 per cent of the people of St. Lucia are descended from black African slaves. This small island country became independent in 1979 after being ruled by Great Britain since 1814.

Fresh fruit is sold at a "floating market" in the harbor at Willemstad on the island of Curaçao. This island, which is the trade and industrial center of the Netherlands Antilles, is dry and barren, so most food must be imported.

A row of palms stand out against a golden sky on an island in the Lesser Antilles. The islands' beautiful landscape and tropical climate attract thousands of vacationers, but many workers must go abroad in search of work because of the lack of job opportunities on the islands.

About A.D. 1000, peaceful farmers called the Arawaks began to arrive. They were followed by the Carib Indians, a warlike people who hunted and fished for a living. The Caribs lived in huts made from palm trees, and they grew corn, pineapples, and tobacco.

Christopher Columbus discovered the West Indies when he landed on an island in the Bahamas in 1492. Over the next 10 years, he claimed almost all the West Indies islands for Spain. Soon, settlers from England, France, the Netherlands, and Denmark—drawn by the promise of gold and other riches—arrived and claimed some of the smaller islands. The newcomers established sugar-cane plantations and imported black slaves from Africa to work on them.

By the late 1800's, slavery was abolished on all the islands. Without cheap labor, the plantation system declined, and the European powers lost interest in their colonies. Since 1945, many islands have become independent or have gained more control over their own affairs.

The island chain of the Lesser Antilles forms the eastern boundary of the Caribbean Sea. On many islands, tourism has become an important part of the economy, but several island governments are working to develop new industries that will lessen their dependence on tourism.

People

The Lesser Antilles have been a popular destination ever since Christopher Columbus discovered what he described as "the best, most fertile, most delightful and most charming land in the world." During the 1500's and 1600's, rival European powers fought for possession of the islands, and today, tourists come from all over the world to bask in the glorious sunshine, swim in the sparkling sea, and stroll along the islands' sandy beaches.

In addition to their great natural beauty, the islands have a charm all their own. Although the Lesser Antilles form a single island group, they vary greatly in culture, government, and language. Each island has kept much of the flavor and characteristics of the European country that once ruled it.

Barbados, for example, earned the nickname "Little England" after more than 300 years of British rule. Although the island country has been independent since 1966, Barbadians are proud that their way of life is very much like that in England. The French dependency of Martinique, on the other hand, has Parisian-style boutiques and imports French wines for its people and for the duty-free tourist shops.

Ethnic origins

The islanders trace their heritage to a variety of sources, from Europe to West Africa, the Orient, and India. Most islanders trace their ancestry to Africans who were captured and taken from their native land to work as slaves on sugar and tobacco plantations. Others have British, Dutch, French, Portuguese, or Spanish ancestors. And some island people are descended from Chinese or East Indian laborers who arrived in the 1800's, after slavery was abolished.

This ethnic mixture has produced a fascinating culture, not only in the Lesser Antilles but throughout the West Indies. The islands' official languages, such as English, Spanish, and Dutch, are often "sprinkled" with African words and phrases. Many islanders use a dialect called *patois*, which is a mixture of African words and mainly English or French. In Aruba and the Netherlands Antilles, the people speak *Papiamento*, which is a combination of Dutch, English, Portuguese, and Spanish.

Steel band musicians in Antigua, *top,* use "pans," or hollowed-out drums, of different heights to achieve a wide range of notes. With 20 or more pans, a modern steel band can play any type of music—from classical to calypso.

Colorful, dramatic costumes, *above,* are part of the fun at carnival, an annual celebration that takes place on most Caribbean islands.

Wide, sandy beaches and graceful palm trees swaying in the ocean breeze lure millions of tourists to the Caribbean islands every year. Vacationers flock to this tropical paradise to escape winter's cold, as well as to enjoy such sports as windsurfing and snorkeling.

Young cricket players enjoy a game under the bright Caribbean sunshine, *bottom center*. The British brought this game to the islands during colonial times.

In Curaçao, a red mailbox gives a glimpse of the island's Dutch heritage. West Indian culture has absorbed many different European, African, and Asian influences, yet it has its own unique style.

Calypso and carnival

The islands of the Caribbean have a number of distinctive traditions, including *calypso*—folk music that originated among the black African slaves on the island of Trinidad. It developed as a way for the slaves to communicate information and feelings, because they were not allowed to speak to one another.

The Caribbean's famous steel band music—or "pan" music, as it is known locally—has its roots in the drums of black Africa. The "pans" are hollow-topped metal drums made from oil drums. The bottom of the oil drum is cut off, and the top is hammered into a concave shape. A characteristic ringing sound is produced when the drum top is struck with a mallet.

Calypso and steel band performances are especially festive during the carnivals that take place on most islands every year. Each island celebrates in its own special way, but colorful parades, along with plenty of food and drink, are always part of the festivities.

The Leeward Islands

Sheltered from the trade winds and boasting a dry, healthful climate, the Leeward Islands stretch from Puerto Rico to the Windward Islands and form part of the Lesser Antilles. With about 15 islands and many islets, the Leeward Islands cover an area of 1,542 square miles (3,994 square kilometers).

St. Martin/Sint Maarten

St. Martin is a hilly island located east of Puerto Rico in the northern part of the arc of the Leeward Islands. The northern section of the island, known as St. Martin, is a dependency of the French overseas department of Guadeloupe. The southern section, known as Sint Maarten, is part of the Netherlands Antilles, a self-governing and equal partner with the Netherlands in the Kingdom of the Netherlands. The governor of Sint Maarten is appointed by the Dutch monarch.

The border between the two areas is an invisible line that visitors often cross without even knowing it. But each side has kept its separate identity, which adds to the island's unique character.

Marigot, the chief city in the northern part of the island, has a distinctly French flavor. Small cafes serve coffee and *croissants* (crescent-shaped rolls), and gourmet restaurants feature freshly caught fish and imported French wines. In the busy Dutch capital of Philipsburg, tourists shop for duty-free china, cosmetics, crystal, jewelry, linens, and perfume imported from the Netherlands and other European countries.

Saba and St. Eustatius

Not far from St. Martin/Sint Maarten are the friendly, slow-paced islands of Saba and St. Eustatius, two small volcanic islands that belong to the Netherlands Antilles. Saba has no beaches, but its dramatic cliffs, deep ravines, and lush tropical vegetation attract many visitors. The capital of Saba, called The Bottom, has all the charm of a Dutch village. The town lies in a green valley that is actually the crater of an extinct volcano.

St. Eustatius, also called Statia, is noted more for its history than its scenery. During the 1700's, St. Eustatius was a neutral free port known as *The Golden Rock,* where mer-

Red-roofed buildings in the colonial style line the tree-shaded central square of Plymouth, the capital of Montserrat. British and Irish settlers colonized the island in 1632, and some islanders still speak with a gentle Irish *brogue* (accent).

Pleasure boats lie at anchor off one of St.-Barthélemy's magnificent beaches, *right.* The beauty of this remote island has attracted many wealthy vacationers, and some have purchased large private estates on the island.

Two sailors on St. Martin/Sint Maarten welcome another sunny day on this charming island. People of many different races live in harmony on the islands of the Caribbean. Most islanders have black African, European, or mixed African and European ancestry.

Clouds cover the rocky mountain peaks on Saba, *left.* Majestic Mount Scenery towers 2,855 feet (870 meters) above this tiny tropical paradise, which is only about 5 square miles (13 square kilometers) in size.

The Leeward Islands include two independent countries: Antigua and Barbuda and St. Christopher and Nevis. The Leewards also include Anguilla, the British Virgin Islands, Montserrat, St. Eustatius, Saba, St. Martin, Guadeloupe, and the U.S. Virgin Islands.

chants, traders, and smugglers bought and sold a variety of goods. St. Eustatius was also the first foreign port in the world to salute the new American flag in 1776.

St.-Barthélemy and Montserrat

The small, mountainous island of St.-Barthélemy, near St. Martin/Sint Maarten, is part of Guadeloupe, a group of islands that forms an overseas department of France. The island's economy is based on farming, fishing, and tourism. St.-Barthélemy is an elegant, sophisticated French island with charming villages and beautiful beaches.

The island was occupied by the French in 1648, but France turned it over to Sweden in 1784. St.-Barthélemy once again came under French rule in 1877. Today, the people still speak an old northern French dialect, and many islanders wear traditional costumes from the French provinces of Brittany and Normandy during festivals and holidays.

Lying southwest of Antigua is Montserrat, a self-governing British crown colony. Montserrat is a tropical island with volcanic landscapes rising to as high as 3,000 feet (914 meters). Farming and tourism are the chief economic activities.

Liberia

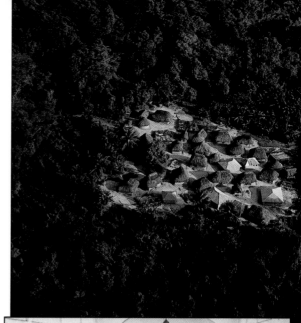

Liberia is Africa's oldest republic. Founded in 1822, it was settled by freed American slaves. The name *Liberia* comes from a Latin phrase meaning *free land*.

Today, almost all of Liberia's people are black Africans, but this population is made up of two groups: the *indigenous* (native) Africans, whose ancestors lived in the region for hundreds of years; and the Americo Liberians, who are descended mainly from African American settlers who emigrated from the United States.

The indigenous Africans, who make up about 95 per cent of the population, belong to 16 different ethnic groups. The largest groups are the Kpelle and the Bassa. The Americo Liberians make up 5 per cent of the population. They live in coastal cities and towns, where the standard of living is higher than in rural areas, and are generally far better off than the indigenous Africans.

Some city dwellers live in expensive homes, but most live in small, wooden houses with tin roofs. However, most people in towns and cities have electricity and running water. Many urban workers are employed in service industries, which have become increasingly important to Liberia's economy. Others work in factories, producing soap, beverages, and explosives.

FACT BOX

COUNTRY

Official name: Republic of Liberia
Capital: Monrovia
Terrain: Mostly flat to rolling coastal plains rising to rolling plateau and low mountains in northeast
Area: 43,000 sq. mi. (111,370 km²)

Climate: Tropical; hot, humid; dry winters with hot days and cool to cold nights; wet, cloudy summers with frequent heavy showers
Main rivers: Cavally, St. Paul
Highest elevation: Nimba Mts., 4,528 ft. (1,380 m)
Lowest elevation: Atlantic Ocean, sea level

GOVERNMENT

Form of government: Republic
Head of state: President
Head of government: President
Administrative areas: 13 counties

Legislature: National Assembly consisting of the Senate with 26 members serving nine-year terms and the House of Representatives with 64 members serving six-year terms
Court system: Supreme Court
Armed forces: N/A

PEOPLE

Estimated 2002 population: 3,385,000
Population growth: 1.94%
Population density: 79 persons per sq. mi. (30 per km²)
Population distribution: 55% rural, 45% urban
Life expectancy in years: Male: 50 Female: 52
Doctors per 1,000 people: 0.0
Percentage of age-appropriate population enrolled in the following educational levels: Primary: 83 Secondary: 24 Further: 7

A Liberian village of thatched-roof mud houses lies amid the dense vegetation of the inland plateau region.

Christian worshipers pray at a church in Monrovia. More than a third of Liberia's people follow Christianity, but most follow traditional African religions.

Liberia is a republic in western Africa. Its capital, Monrovia, was named for James Monroe, the U.S. president who arranged for its settlement.

Rural Liberians, by contrast, live in villages of mud houses with thatched roofs, and almost none have electricity or indoor plumbing. Most rural Liberians are farmers. Their food crops include cassava, rice, sugar cane, and tropical fruits.

The differences between indigenous Africans and Americo Liberians have created difficulties throughout Liberia's history. While the ancestors of the indigenous Africans probably came from Sudanese kingdoms between the 1100's and 1500's, ancestors of the Americo Liberians—mainly freed slaves—began to settle in the region in 1822. Not only did the settlers have trouble finding food, but they often fought with the indigenous Africans over land.

In 1838, the Commonwealth of Liberia was formed, but it was still controlled by the American Colonization Society, a group of white Americans who had helped the former slaves settle the area. Liberia became independent on July 26, 1847.

The Liberian economy was helped immensely in 1926, when the government leased large amounts of land to the American Firestone Company for rubber plantations. It was helped still more in 1944, when William V. Tubman, an Americo Liberian, became president. Tubman, determined to improve the nation's economy, expanded mining and foreign trade.

Tubman was succeeded by William R. Tolbert, Jr., another Americo Liberian, but under Tolbert, the rich prospered while the poor became poorer. In 1980, a group of indigenous Liberian army officers killed Tolbert and took control.

Samuel K. Doe, an army sergeant, was elected president in 1985. However, many people felt he had fixed the election. Under Doe, Liberia faced major economic problems. Doe also had many of his political opponents jailed or killed. A bloody civil war broke out in the late 1980's, and in 1990, Doe was captured by rebel forces, tortured, and killed. Later that year, a cease-fire went into effect.

In October 1992, fighting between the new government and opponents broke out. A peace pact signed in July 1993 failed to end the civil war. In September 1994, leaders of three major factions signed a new treaty. They agreed to share power until October 1995, and to hold elections within a year. Fighting continued until 1996, when factional leaders agreed to a cease-fire and drafted a peace plan. In 1997, Liberia ended its seven-year civil war, held its first free elections since 1971, and inaugurated a new civilian government.

ECONOMY

Languages spoken:
English 20% (official) some 20 ethnic group languages

Religions:
indigenous beliefs 40%
Christian 40%
Muslim 20%

TECHNOLOGY

Radios per 1,000 people: 274

Televisions per 1,000 people: 25

Computers per 1,000 people: N/A

Currency: Liberian dollar

Gross domestic product (GDP) in 2001: $3.6 billion U.S.

Real annual growth rate (2001): 0.5%

GDP per capita (2001): $1,100 U.S.

Balance of payments (2001): N/A

Goods exported: Diamonds, iron ore, rubber, timber, coffee, cocoa

Goods imported: Fuels, chemicals, machinery, transportation equipment, manufactured goods; rice and other foodstuffs

Trading partners: South Korea, Benelux, Norway, Ukraine, Japan, Italy

Her face painted to ward off spirits, a girl from Liberia's coastal region carries a tray of fish, *above*. Most of Liberia's people come from native African groups. The descendants of freed slaves account for only 5 per cent of the population.

827

Libya

Libya, an Arab country in North Africa, forms a kind of bridge between Islamic countries. To the west are the *Maghreb* (western) countries of Tunisia, Algeria, and Morocco, which were colonized by France. To the east lies the Middle East—Egypt and the countries of the Arabian Peninsula.

Government

The government of Libya is based on popular assemblies. About 1,000 local groups elect representatives to the General People's Congress (GPC), which officially runs the government. The GPC meets every year to pass laws and select the members of the General People's Committee, which forms national policy. Libya is divided into 25 *baladiyat.* Each baldiya is run by a local People's Congress. All Libyan citizens 18 years old or older may vote and hold office.

Although Libya is a republic, much control lies in the hands of Colonel Muammar Muhammad al-Qadhafi. Qadhafi is Libya's head of state, but he has no official title. Qadhafi took over after he led a revolt against Libya's King Idris in 1969.

Under Qadhafi's strong rule, Libya's government has taken control of most of the country's economic activities. Oil revenues have been used to fund new and ambitious social, political, and economic projects.

Qadhafi replaced the political institutions of Libya with the popular assemblies governing the country today. Yet, despite the appearance of democracy in Libya, Qadhafi tolerates no political opposition. Although the GPC technically chooses the members of the General People's Committee, Qadhafi actually controls these appointments. Political parties have been forbidden in Libya since 1952. In 1971, Qadhafi formed the Arab Socialist Union as Libya's only political alliance. However, a number of underground opposition groups exist.

History

Until the early 1900's, Libya consisted of three separate regions—each with its own distinctive geography and history. In the 600's B.C., Greeks colonized the northeastern coast and established a province known as Cyrenaica. In the 400's B.C., the Carthaginians built trading centers on the northwestern coast; this region was called Tripolitania. Nomads lived in the southwestern desert known as the Fezzan.

Over the next 1,000 years, different empires took control of one or more of these Libyan regions. Then in A.D. 642, the Arabs invaded Cyrenaica; in 643, they occupied the city of Tripoli. The Arabs brought their language and their Islamic religion to Libya.

Turks from the Ottoman Empire captured all three regions in the 1500's and controlled

FACT BOX

LIBYA

COUNTRY

Official name: Al Jumahiriyah al Arabiyah al Libiyah ash Shabiyah al Ishtirakiyah (Socialist People's Libyan Arab Jamahiriya)
Capital: Tripoli
Terrain: Mostly barren, flat to undulating plains, plateaus, depressions

Area: 679,362 sq. mi. (1,759,540 km²)
Climate: Mediterranean along coast; dry, extreme desert interior
Highest elevation: Bette Peak, 7,500 ft. (2,286 m)
Lowest elevation: Sabkhat Ghuzayyil, 154 ft. (47 m) below sea level

GOVERNMENT

Form of government: Jamahiriya (a state of the masses)
Head of state: no official title for head of state
Head of government: Secretary of the General People's Committee (Premier)

Administrative areas: 25 baladiyat (municipalities)
Legislature: General People's Congress
Court system: Supreme Court
Armed forces: 65,000 troops

PEOPLE

Estimated 2002 population: 5,869,000
Population growth: 2.42%
Population density: 9 persons per sq. mi. (3 per km²)
Population distribution: 86% urban, 14% rural
Life expectancy in years: Male: 73 Female: 78
Doctors per 1,000 people: 1.3
Percentage of age-appropriate population enrolled in the following educational levels: Primary: 153* Secondary: 77 Further: 57

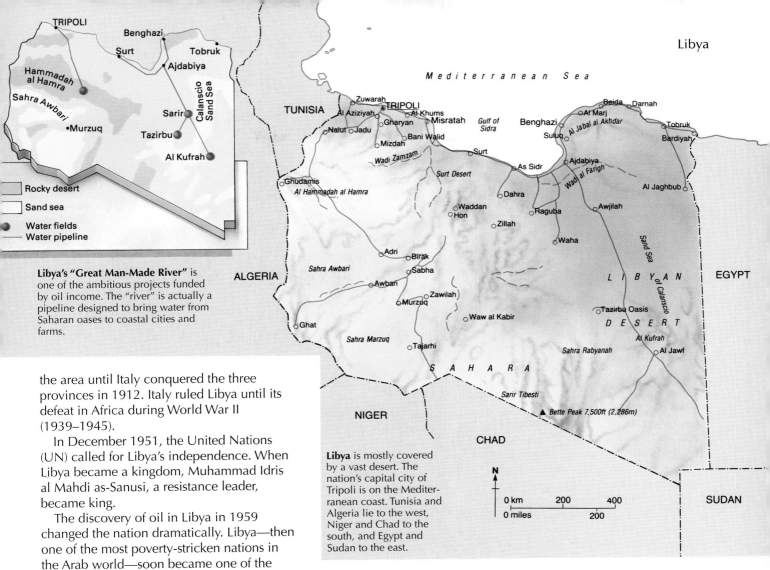

Mediterranean Sea

Libya's "Great Man-Made River" is one of the ambitious projects funded by oil income. The "river" is actually a pipeline designed to bring water from Saharan oases to coastal cities and farms.

Rocky desert
Sand sea
● Water fields
— Water pipeline

Libya is mostly covered by a vast desert. The nation's capital city of Tripoli is on the Mediterranean coast. Tunisia and Algeria lie to the west, Niger and Chad to the south, and Egypt and Sudan to the east.

0 km   200   400
0 miles   200

the area until Italy conquered the three provinces in 1912. Italy ruled Libya until its defeat in Africa during World War II (1939–1945).

In December 1951, the United Nations (UN) called for Libya's independence. When Libya became a kingdom, Muhammad Idris al Mahdi as-Sanusi, a resistance leader, became king.

The discovery of oil in Libya in 1959 changed the nation dramatically. Libya—then one of the most poverty-stricken nations in the Arab world—soon became one of the wealthiest. However, many of Libya's people

became discontented because so few of them shared in the oil wealth.

In September 1969, a group of military officers called the Revolutionary Command Council overthrew King Idris, and Qadhafi, its leader, rose to power. Qadhafi tried to forge unions with other Arab states, but none of his efforts succeeded. During the 1970's, Libya supported radical movements around the world, particularly the Palestine Liberation Organization (PLO), and aided rebellions in Chad and Morocco.

The leaders of many nations have criticized Qadhafi for interfering in the affairs of other countries. In 1986 and again in 1989, United States planes attacked Libyan military bases or aircraft. In 1992, the UN imposed economic sanctions on Libya for harboring two Libyans suspected of bombing a U.S. airplane in 1988 and a French airplane in 1989, killing 441 people. In 1994, Qadhafi celebrated 25 years in power.

Languages spoken:
Arabic
Italian
English
Religion:
Sunni Muslim 97%

Enrollment ratios compare the number of students enrolled to the population which, by age, should be enrolled. A ratio higher than 100 indicates that students older or younger than the typical age range are also enrolled

TECHNOLOGY

Radios per 1,000 people: 273
Televisions per 1,000 people: 137
Computers per 1,000 people: N/A

ECONOMY

Currency: Libyan dinar
Gross domestic product (GDP) in 2001: $40 billion U.S.
Real annual growth rate (2001): 3%
GDP per capita (2001): $7,600 U.S.
Balance of payments (2000): $1,984
Goods exported: Crude oil, refined petroleum products, natural gas
Goods imported: Machinery, transport equipment, food, manufactured goods
Trading partners: Italy, Germany, United Kingdom, Spain

Muammar Muhammad al-Qadhafi, Libya's head of state, took control in 1969. Qadhafi's *The Green Book,* published in the mid-1970's, outlines his program for Libya's political, social, and economic development.

Land and Economy

The Sahara dominates the landscape of Libya, covering about 95 per cent of the country. In eastern Libya, the Sahara is called the Libyan Desert.

Huge sand dunes make up most of the Sahara in Libya. The desert terrain gradually rises from north to south, where rugged mountains rise from the desert floor along the southern border. Majestic Bette Peak, Libya's highest point, towers 7,500 feet (2,286 meters) above the desert sands.

The desert has extremely hot daytime temperatures but cools off rapidly at night. The average daytime temperature is 100° F. (38° C); the average nighttime temperature is 50° F. (10° C). An average of less than 2 inches (5 centimeters) of rain falls each year.

Except for scattered oases, only the land near the Mediterranean Sea in northwestern and northeastern Libya is inhabitable. In north-central Libya, the Sahara reaches to the coast.

Most Libyans live on plains near the Mediterranean Sea. One fertile strip of land stretches along the northwest coast for almost 200 miles (320 kilometers). It is rimmed by hills to the south. Other Libyans live on a smaller plain on the northeast coast between Benghazi and Darnah, or in the Green Mountain region (Al Jabal al Akhdar)—the highlands to the south.

Both coastal areas have a Mediterranean climate with warm summers and mild winters. January temperatures in Tripoli, on the northwest coast, average 52° F. (11° C); July temperatures average 81° F. (27° C). The coast receives more rain than the inland desert—about 16 inches (41 centimeters) per year.

Most Libyan workers are employed in service industries or farm in the coastal areas. But Libya's most valuable source of income—the oil fields that form the basis of the Libyan economy—is centered in the north-central region, where the Sahara reaches to the Gulf of Sidra.

Petroleum accounts for about 50 per cent of the value of Libya's economic production, and for almost all its export earnings. The government, which controls petroleum operations, has used some of the oil wealth to improve farmland and provide services for

Rocky plateaus and gravelly plains are typical of most of Libya's landscape, *right*. The Sahara covers more than 90 per cent of the country. The road cutting through the desert in this photo serves an oil field.

The Tuareg, a nomadic Berber group, gather feed for their camels. The Tuareg and other nomads roam across the desert in search of water and grazing land for their animals. Scattered oases are the only fertile areas in the desert of Libya.

the Libyan people, and to pay for imported manufactured goods, food, technology, and weapons.

Service industries make up about 32 per cent of the value of Libya's economic production. Government agencies rank first among service industries, which also include banks, retail stores, and wholesale companies.

Construction accounts for about 12 per cent of the value of the country's economic production. Construction boomed after 1959, when many factories and other buildings were needed for the rapidly developing petroleum industry.

Manufacturing makes up about 4 per cent of the value of Libya's economic production, with refined petroleum products and petrochemicals ranking first. Other goods include cement, processed foods, and steel. Libya's northern cities are the chief manufacturing centers.

Although agriculture accounts for only about 2 per cent of the value of economic production in Libya today, about 18 per cent of Libya's people are farmers. The main crops include barley, citrus fruits, dates, olives, potatoes, tomatoes, and wheat. Farmers and herders also raise cattle, chickens, and sheep. Libya imports most of its food, however, because only about 5 per cent of the land can be farmed.

Libya's national airline, Libyan Arab Airlines, connects Libya with other countries. The country's major seaports are Tripoli, Benghazi, Misratah, and Port Brega. Libya has no railroads, but paved roads link the larger cities of northern Libya and connect them with desert oases. About 10 per cent of the people own automobiles.

Camels and donkeys have served as beasts of burden in the Sahara for centuries. Modern Libya, however, has a developing economy that depends heavily on petroleum, and 70 per cent of its people live in urban areas.

A silversmith works at a traditional Libyan craft. Such crafts are declining today because fewer young people are learning these ancient skills. Young Libyans are increasingly attracted to higher-paid work, such as jobs in the oil industry, which employs a small percentage of Libya's work force.

Economy

Land use in Libya consists mostly of nomadic herding and petroleum-related activities, such as the recovering and refining of oil and natural gas. Only small areas near the coast are suitable for farming.

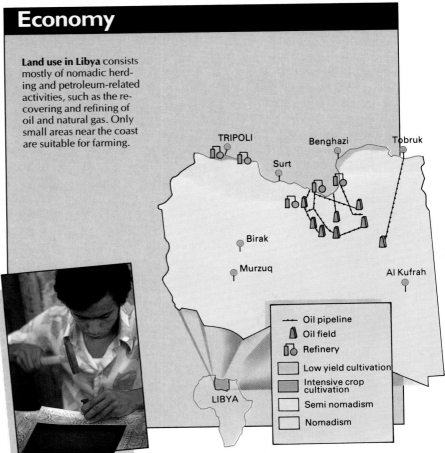

TRIPOLI
Surt
Benghazi
Tobruk
Birak
Murzuq
Al Kufrah
LIBYA

- Oil pipeline
- Oil field
- Refinery
- Low yield cultivation
- Intensive crop cultivation
- Semi nomadism
- Nomadism

People

More than 5 million people live in Libya. About 80 per cent of the population live along the Mediterranean coast or in the upland regions just south of it.

The early inhabitants of Libya were Berbers who began moving into North Africa about 3000 B.C., probably from southwestern Asia or Europe. Beginning in the 600's B.C., parts of Libya were then colonized or invaded by a series of forces: the Carthaginians, the Greeks, the Romans, the Vandals, and the Byzantines. In the A.D. 600's, Arabs invaded first northeastern and then northwestern Libya. Their culture, language, and religion were adopted by the native Berbers, and today, more than 90 per cent of the Libyan people are of mixed Arab and Berber ancestry. Although Arabic is the official language of Libya, many educated Libyans also speak English or Italian.

Islam is the official religion of Libya. According to constitutions written in 1977, all Libyan laws must agree with Islamic teachings. Almost all Libyans are *Sunni Muslims*—that is, followers of Islam who belong to the Sunni branch of that faith.

About 85 per cent of Libya's people live in cities. About 15 per cent live in villages or oases, and a small number are nomads who move across the desert in search of water and pasture for their sheep, goats, and camels.

Today, Libya is an urban society. Since the nation's economy, spurred by the oil industry, started to develop in the mid-1900's, many rural Libyans have moved to urban areas. The flood of newcomers has caused serious overcrowding in the cities, especially in the older neighborhoods.

This shift in population has brought many changes in the Libyan way of life. Rural Libyans, for example, live in extended families, with grandparents, parents, and children sharing a home. This living arrangement is seldom followed in the crowded cities, however.

In rural areas, most people live in stone or mud-brick houses that often have only one room for living, cooking, eating, and sleeping. The animals are kept in a nearby shelter. In the cities, high-rise apartment buildings as well as office buildings line the downtown streets. In fact, Libya's large cities look much like those in North America and Western Europe. In the suburbs, some Libyans live in large single-family homes.

Most rural Libyans wear traditional clothing. The men wear a loose cotton shirt and trousers covered by an outer cloak, and a flat, brimless, tight-fitting cap. The women wear a full-length robe. Urban Libyans, on the other hand, generally wear Western-style clothing. However, some wear traditional dress to show their respect for time-honored values and practices.

The status of Libyan women has also changed dramatically since the mid-1900's.

Muslims enter the Kara-manli mosque in Tripoli, *left,* through a colon-naded passageway. Qad-hafi's government has promoted traditional Islamic practices along with modern social reforms, such as increased freedom for women.

Tripoli's fine harbor—a stronghold for pirates in the 1700's—makes the city a trading center for farmers who live in the region. Nearly a million Libyans live in the capital city of Tripoli. The Arabic name for Tripoli is Tarabulus.

The ruins of a Roman theater, *below center,* still stand at Sabrata on Libya's Mediterranean coast, a reminder of the days when northwestern Libya was part of the Roman province of Africa Nova.

Whitewashed mud-brick houses, such as the one shown below, are the typical homes of Libya's rural population. Libya's city dwellers, on the other hand, live in modern high-rise apartments.

In the past, women in Libya received little or no education and were largely confined to their homes. Today, Libyan women have the legal right to participate fully in society. Although many traditional attitudes toward women remain, women make up 10 per cent of the nation's work force today—and this percentage is increasing as more women receive an education.

About 75 per cent of Libya's people can read and write. This literacy rate is largely due to the government's education program, which is funded by oil income.

The Sahara

Libya has the distinction of being the location of the world's highest official temperature. In 1922, at Al Aziziyah in the Sahara, the temperature reached 136° F. (58° C). For the Libyan people, the harsh Saharan climate and landscape are a fact of life. But the desert itself is a unique environment where plants and animals have adapted to the extreme conditions.

The Sahara is by far the world's largest desert. It covers about 3-1/2 million square miles (9 million square kilometers) of North Africa, stretching more than 3,500 miles (5,630 kilometers) from west to east and 1,200 miles (1,930 kilometers) from north to south.

The name *Sahara* comes from the Arabic word *sahra*, which means *desert*. The dry, hot climate of the Sahara is typical of many deserts. The average yearly rainfall is less than 8 inches (20 centimeters), but large areas receive less than 1 inch (2.5 centimeters) of rain per year.

In the central portion of the Sahara, mountains and uplands rise as high as 11,204 feet (3,415 meters). This highland region gets slightly more rain than other areas of the desert, and snow may even cover some peaks in the winter.

Most of the Sahara is made up of rocky plateaus and gravelly plains, but lying within large basins in the desert are vast seas of sand called *ergs*. The shifting sands in these ergs sometimes form towering dunes as high as 600 feet (180 meters).

The plateaus and plains of the Sahara are not entirely barren, however. Plant and animal life does exist, though it may not be as plentiful as in other deserts.

The grasses, shrubs, and trees that grow in parts of the Sahara have adapted to the arid conditions there. Some of the plants are *ephemeral*, or short-lived. Their seeds lie in the ground and do not start to grow until rain falls. Then the plants shoot up quickly and may complete their life cycle in just six to eight weeks.

Plants that live longer than a year have developed various ways to get—and keep—water. Some have long roots that reach deep into the soil and absorb moisture; others take in moisture from the air through their leaves.

Animals also have adapted to the arid Sahara. Barbary sheep live on the rocky plateaus of the desert, while white gazelles and rare antelope called *addax* roam the sand dunes. Horned vipers, spiny-tailed lizards, gerbils, and fennec foxes also live in the dunes. Many of the small animals are *nocturnal* animals, which stay in their burrows during the day to avoid the heat, emerging at night to hunt for food.

Camels were introduced into the Sahara about 2,000 years ago. Like most desert animals, they can go for long periods without drinking water. They get some water from the

Saharan wildlife
has adapted to life in the desert. Larger Saharan animals include the dorcas gazelle (1), the addax (2), and the oryx (3). Small rodents, such as the jerboa (4), have specially adapted kidneys that reduce their water loss. The horned viper (5) moves in a sidewinding motion across the desert sand.

The desert locust (6) usually travels in swarms, while the dung beetle (7) burrows underground. It lays its eggs in the manure of large animals. When the sand grouse (8) finds water, it soaks its breast feathers so that it can carry the moisture back to its young. The large ears

plants they eat, and they keep water in their bodies because they do not sweat much.

Scattered throughout the Sahara are *oases*—fertile green spots irrigated by water from springs or wells. Plants and animals that require more water—such as huge date palms—thrive in the oases. Some oases are large enough to support villages of up to 2,000 people.

Thousands of years ago, the Saharan environment had a much wetter climate than it has now. Grasslands covered much of the region, and some areas had lakes and streams. Elephants, giraffes, and other large animals roamed the land.

About 4000 B.C., Africa's climate started to become drier, and the Sahara region began to turn into a desert. Ever since, the desert has been spreading. Through the centuries, people have contributed to this expansion by removing the plants that help to hold down the soil against erosion. The farmers have allowed their herds to overgraze on grasses at the edges of the desert, and they have cut down trees and shrubs there for firewood.

and long legs of the nocturnal desert hedgehog (9) help it lose body heat. The skink (10)—a smooth-scaled lizard—"swims" through the desert sand just below the surface, hunting for insects. The fennec (11), a type of fox, hunts small animals and insects at night.

Rock-strewn plateaus and gravelly plains make up most of the Sahara. Formations of granite, sandstone, and other types of rock, sculpted by wind and water, rise from the desert floor throughout the Sahara.

Liechtenstein

Liechtenstein, in south-central Europe, is one of the smallest countries in the world, with an area of only 62 square miles (160 square kilometers) and a total population of 32,000. This tiny scenic land lies along the Rhine River between Switzerland and Austria. Vaduz, with almost 5,000 people, is the country's capital city.

Liechtenstein is a constitutional monarchy, ruled by a prince—the head of the House of Liechtenstein. The 25 members of the country's *Landtag* (parliament) pass its laws, which must, however, be approved by the prince. A five-member Collegial Board headed by a prime minister handles government operations.

The people of Liechtenstein have many close cultural, political, and economic ties with Switzerland. Liechtenstein uses the Swiss franc, and Switzerland not only operates the country's communication systems but also represents Liechtenstein in its diplomatic and trade relations.

Economy

Since about 1950, Liechtenstein has been transformed from an agricultural country to a highly industrialized one. Today, it has one of the world's highest standards of living.

Products made in Liechtenstein include ceramics, cotton textiles, appliances, pharmaceuticals, and precision instruments.

The landscape of lovely Liechtenstein, *right,* ranges from rich farmland along the Rhine Valley to snow-topped mountains and alpine meadows.

The prince of Liechtenstein's art collection, *far right, above,* includes works by Rembrandt and Botticelli and ranks as one of the world's finest private art collections. Most of these artworks are housed in the prince's castle in Vaduz.

Many citizens dress in colorful traditional costumes, *far right, below,* to celebrate Liechtenstein's independence.

FACT BOX

COUNTRY

Official name: Fuerstentum Liechtenstein (Principality of Liechtenstein)
Capital: Vaduz
Terrain: Mostly mountainous (Alps) with Rhine Valley in western third
Area: 62 sq. mi. (160 km²)

Climate: Continental; cold, cloudy winters with frequent snow or rain; cool to moderately warm, cloudy, humid summers
Main rivers: Rhine, Samina
Highest elevation: Grauspitz, 8,527 ft. (2,599 m)
Lowest elevation: Ruggeller Riet, 1,411 ft. (430 m)

GOVERNMENT

Form of government: Hereditary constitutional monarchy
Head of state: Monarch
Head of government: Head of government and Deputy Head of Government
Administrative areas: 11 Gemeinden (communes)

Legislature: Landtag (Diet) with 25 members serving four-year terms
Court system: Oberster Gerichtshof (Supreme Court), Obergericht (Superior Court)
Armed forces: Switzerland is responsible for Liechtenstein's defense

PEOPLE

Estimated 2002 population: 34,000
Population growth: 1.02%
Population density: 548 persons per sq. mi. (213 per km²)
Population distribution: 80% rural, 20% urban
Life expectancy in years: Male: 75 Female: 82
Doctors per 1,000 people: N/A
Percentage of age-appropriate population enrolled in the following educational levels: Primary: N/A Secondary: N/A Further: N/A

Liechtenstein is a tiny country in south-central Europe.

Agricultural products include wheat and potatoes grown in the Rhine Valley, and farmers also raise beef and dairy cattle on the grassy meadows of the country's mountain slopes. Grapes and other fruits are grown on the upland slopes. Liechtenstein's low tax rates attract many foreign businesses, and the government collects money from more than 5,000 foreign firms now based in Liechtenstein.

History

Charlemagne controlled the area that is now Liechtenstein in the late 700's and early 800's. After his death, the area was divided into two independent states. Both states later became part of the Holy Roman Empire. By 1712, both states had been acquired by Johann-Adam Liechtenstein, a prince from Vienna, whose descendants still rule Liechtenstein today.

The country has been an independent state since 1719, except for a brief period during the early 1800's when Napoleon controlled it. Liechtenstein has also been a neutral country since 1866 and has had no army since 1868.

The country had strong economic ties with Austria from 1852 until after World War I ended in 1918. In 1924, Liechtenstein agreed to an economic union with Switzerland. The women of Liechtenstein were finally given the right to vote in 1984.

Languages spoken:
German (official)
Alemannic dialect

Religions:
Roman Catholic 80%
Protestant 7%

TECHNOLOGY

Radios per 1,000 people: N/A

Televisions per 1,000 people: N/A

Computers per 1,000 people: N/A

ECONOMY

Currency: Swiss franc, franken, or franco

Gross domestic product (GDP) in 1998: $730 million U.S.

Real annual growth rate (2000): N/A

GDP per capita (1998): $23,000 U.S.

Balance of payments (2000): N/A

Goods exported: Small specialty machinery, dental products, stamps, hardware, pottery

Goods imported: Machinery, metal goods, textiles, foodstuffs, motor vehicles

Trading partners: European Union, European Free Trade Association, Switzerland

Lithuania

The largest of the Baltic States, Lithuania is situated on the eastern shore of the Baltic Sea, south of Latvia. From 1940 until 1991, Lithuania was a Soviet republic, under the control of the Soviet central government.

Lithuania's landscape consists chiefly of flat or gently sloping land with some 3,000 small lakes covering about 1-1/2 per cent of the region. Beautiful white sand dunes line the coast of the Baltic Sea and stretch for miles along a sparkling strip of land that separates the Kuršiu Marios (a lagoon) from the sea.

About 80 per cent of the people of Lithuania are Lithuanians, a nationality group that has its own customs and language. About 9 per cent of the population are Russians, and a somewhat smaller number are Poles.

History

The ancestors of present-day Lithuanians lived in the region about the time of Christ. Near the end of the 1100's, these people united into a single nation.

During the Middle Ages, the people of Lithuania controlled an empire that extended to the Black Sea in the south and almost as far as Moscow in the east. In 1386, Grand Duke Jagiełło joined Lithuania with Poland in a union that lasted more than 400 years, until the czar of Russia gained control of the Lithuanian part of the region in 1795.

In 1918—during World War I—Lithuania declared its independence. Although Russia attempted to take over the country, it was

FACT BOX

LITHUANIA

COUNTRY

Official name: Lietuvos Respublika (Republic of Lithuania)
Capital: Vilnius
Terrain: Lowland, many scattered small lakes, fertile soil
Area: 25,174 sq. mi. (65,200 km²)

Climate: Transitional, between maritime and continental; wet, moderate winters and summers
Main river: Nemunas
Highest elevation: Juozapines/Kalnas, 958 ft. (292 m)
Lowest elevation: Baltic Sea, sea level

GOVERNMENT

Form of government: Parliamentary democracy
Head of state: President
Head of government: Premier
Administrative areas: 44 rajonai (regions), 11 municipalities

Legislature: Seimas (Parliament) with 141 members serving four-year terms
Court system: Supreme Court, Court of Appeal
Armed forces: 12,130 troops

PEOPLE

Estimated 2002 population: 3,647,000
Population growth: -0.29%
Population density: 145 persons per sq. mi. (56 per km²)
Population distribution: 68% urban, 32% rural
Life expectancy in years:
Male: 63
Female: 75
Doctors per 1,000 people: 4.0
Percentage of age-appropriate population enrolled in the following educational levels:
Primary: 101*
Secondary: 90
Further: 4
Languages spoken:
Lithuanian (official)
Polish
Russian

Vilnius, Lithuania's capital and largest city, is an important industrial, transportation, and cultural center. Vilnius is best known for its many impressive churches and other buildings dating from the 1400's.

Forests and lakes cover much of Lithuania's landscape, but industrial pollution now seriously threatens the entire region.

The largest of the three Baltic States, which also include Estonia and Latvia, Lithuania, *map below right,* lies in northern Europe along the eastern shore of the Baltic Sea.

defeated in 1920. Poland occupied the Lithuanian capital of Vilnius between 1920 and 1939, and in 1940, the Soviet Union annexed the country.

Resistance and independence

The Lithuanians strongly opposed the Soviet take-over. In the late 1980's, Lithuanian nationalism gained massive support throughout the republic, as leaders of the non-Communist *Sąjūdis* and other groups sought to restore political and economic independence.

In August 1991, conservative Communist leaders in the Soviet Union attempted to overthrow the central government, but failed. In the political upheaval that followed, Lithuania declared its independence from the Soviet Union. In September 1991, the Soviet government recognized Lithuania's independence. Afterward, the government of Lithuania expressed support of free enterprise and initiated plans to establish a free market economy. In 1994, Lithuania joined the Partnership for Peace, a North Atlantic Treaty Organization (NATO) program of cooperation with former members of the Warsaw Pact, a Soviet military alliance.

Religions:
Roman Catholic
(primarily)
Lutheran
Russian Orthodox
Protestant
evangelical Christian
Baptist
Muslim
Jewish

Enrollment ratios compare the number of students enrolled to the population which, by age, should be enrolled. A ratio higher than 100 indicates that students older or younger than the typical age range are also enrolled.

TECHNOLOGY

Radios per 1,000 people:
500

Televisions per 1,000 people: 422

Computers per 1,000 people: 64.9

ECONOMY

Currency: Lithuanian litas

Gross national income (GNI) in 2000: $10.8 billion U.S.

Real annual growth rate (1999–2000): 3.9%

GNI per capita (2000): $2,930 U.S.

Balance of payments (2000): -$675 million U.S.

Goods exported: Machinery and equipment, mineral products, textiles and clothing, chemicals, foodstuffs

Goods imported: Machinery and equipment, mineral products, chemicals, textiles and clothing, foodstuffs

Trading partners: Russia, Germany, Latvia, Denmark

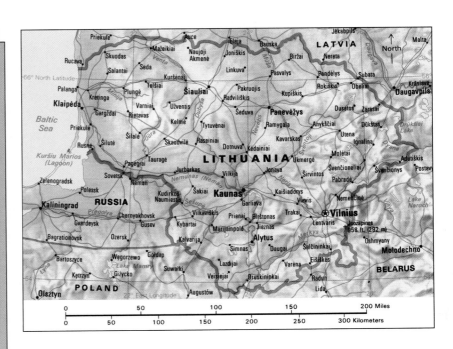

Luxembourg

Luxembourg, one of Europe's oldest and smallest independent countries, lies in northwestern Europe where Germany, France, and Belgium meet. The country covers only 998 square miles (2,586 square kilometers) and has about 438,000 people.

The people of Luxembourg often call their country the "green heart of Europe" because of its beautiful landscape. Rolling green hills and medieval castles perched on steep cliffs attract many tourists.

The northern third of Luxembourg is a wooded region—an extension of the Ardennes mountain system of Belgium and Germany. River valleys cut through the region's low hills. The Bon Pays (Good Land) region, which covers the rest of the country, is mainly a hilly or rolling plateau with level areas along its rivers. The Bon Pays is a productive farming region.

Government

Luxembourg is a constitutional monarchy. The grand duke or duchess of the House of Nassau serves as monarch and head of state, but the office is largely ceremonial. The monarchy is inherited by the monarch's oldest son or daughter.

A 60-member parliament called the Chamber of Deputies passes Luxembourg's laws. The people elect the members to five-year terms. With the support of parliament, the monarch appoints a prime minister and 10 other Cabinet ministers to run the government. The monarch also appoints for life a 21-member advisory body called the Council of State.

History

Luxembourg became an independent state in A.D. 963, when Siegfried, Count of Ardennes, gained control of the area and built a castle on the site of the present-day city of Luxembourg. In 1354, Charles IV created the Duchy of Luxembourg, and a period of prosperity began.

The country later came under a succession of foreign rulers, including Burgundy, Spain, Austria, and France. In 1815, after the defeat of Napoleon I of France, the Congress of Vienna made Luxembourg a grand duchy that was technically ruled by the Netherlands. In 1890, Luxembourg broke away from the Netherlands and named its own monarch.

FACT BOX

LUXEMBOURG

COUNTRY

Official name: Grand-Duche de Luxembourg (Grand Duchy of Luxembourg)
Capital: Luxembourg
Terrain: Mostly gently rolling uplands with broad, shallow valleys; uplands to slightly mountainous in the north; steep slope down to Moselle flood plain in the southeast

Area: 998 sq. mi. (2,586 km²)
Climate: Modified continental with mild winters, cool summers
Main rivers: Attert, Allzette, Moselle, Sûre
Highest elevation: Buurgplatz, 1,835 ft. (559 m)
Lowest elevation: Moselle River, 436 ft. (133 m)

GOVERNMENT

Form of government: Constitutional monarchy
Head of state: Grand duke
Head of government: Prime minister
Administrative areas: 3 districts
Legislature: Chambre des Deputes (Chamber of Deputies) with 60 members serving five-year terms
Court system: Cour Superieure de Justice (Superior Court of Justice), Tribunale Administratin (Administrative Court)
Armed forces: 768 troops

PEOPLE

Estimated 2002 population: 438,000
Population growth: 1.27%
Population density: 439 persons per sq. mi. (169 per km²)
Population distribution: 88% urban, 12% rural
Life expectancy in years: Male: 74 Female: 81
Doctors per 1,000 people: N/A
Percentage of age-appropriate population enrolled in the following educational levels: Primary: N/A Secondary: N/A Further: N/A

The European Court of Justice meets in the European Center in the city of Luxembourg. Several other international organizations also have their headquarters here.

Luxembourg, *above,* is one of Europe's smallest countries. Its northern region is hilly and wooded, while its southern two-thirds has most of the nation's farmland and industry.

Languages spoken:
  Luxembourgian
  German
  French
  English
Religions:
  Roman Catholic 97%
  Protestant
  Jewish

TECHNOLOGY

Radios per 1,000 people:
  N/A
Televisions per 1,000 people: N/A
Computers per 1,000 people: N/A

ECONOMY

Currency: Euro
Gross national income (GNI) in 2000:
  $18,439 million U.S.
Real annual growth rate (1999–2000):
  8.5%
GNI per capita (2000): $42,060 U.S.
Balance of payments (2000): N/A
Goods exported: Finished steel products, chemicals, rubber products, glass, aluminum, other industrial products
Goods imported: Minerals, metals, foodstuffs, quality consumer goods
Trading partners: Germany, Belgium, France

Germany occupied the country during parts of World War I (1914–1918) and World War II (1939–1945). In the winter of 1944 and 1945, part of the Battle of the Bulge was fought in northern Luxembourg.

After World War II, Luxembourg became one of Europe's most prosperous countries. In 1945, it became a member of the United Nations, and in 1948, with Belgium and the Netherlands, Luxembourg helped establish the economic union of Benelux. It was also a founding member of the North Atlantic Treaty Organization in 1949 and of the European Economic Community, now known as the European Union, in 1957.

Today, Luxembourg plays an important role in European affairs. Its capital and largest city, also called Luxembourg, serves as the headquarters of the European Court of Justice and several other international agencies. The country is also an international financial center where foreign banks and corporations have offices.

People and Economy

The population of Luxembourg is unevenly spread throughout the country. The most densely populated areas include the capital city of Luxembourg, the industrialized southwestern corner of the country, and the farming regions of the Bon Pays.

More than 95 per cent of Luxembourg's people are Roman Catholics, and most of the population is of German descent. Foreigners, particularly migrant workers and civil servants from all over Europe, make up about 25 per cent of the population.

Luxembourg has three official languages—French, German, and Letzeburgesch, a German dialect that became the third official language in 1985. Most people speak Letzeburgesch, though they usually write in French or German. German is used in most elementary schools and French is used in most high schools. Luxembourg's newspapers are printed in German, while French is used in the courts and parliament.

Luxembourgers have close cultural ties with their neighbors, but their independent spirit is expressed in the words of their national anthem, *"Mir welle bleiwe wat mir sin"* (*"We want to remain what we are"*). Most Luxembourgers enjoy a high standard of living and have better food and housing than many other Europeans have. In addition, the government provides extensive social security and health-care benefits.

Economy

Luxembourg is one of Europe's leading steel producers, but since World War II, the mines that provided the iron ore to make the steel have decreased in number. This reduction in the number of mines plus the general decline of steel production in the 1970's led to the closing of many steel factories. As a result, Luxembourg has diminished its dependence on heavy industry.

Today, Luxembourg has a number of high-technology industries that produce electronic equipment, chemicals, plastics, and tires. Since the 1960's, Luxembourg has also developed many small- and medium-sized service industries, including

tourism—now another major industry. The country has also become an international banking center.

Although about 50 per cent of the nation's land is farmed, only 6 per cent of Luxembourg's workers are farmers. Barley, oats, potatoes, wheat, and livestock are raised in the Bon Pays region. The area along the Moselle River yields grapes that produce fine wine.

The city of Luxembourg

Many of the foreign banks in Luxembourg are based in the country's capital city. Luxembourg also serves as the headquarters for many European international agencies.

The capital occupies a picturesque position on a rocky plateau overlooking the valleys formed by the Alzette and Petrusse rivers. The Grand Ducal Palace in Luxembourg, built in the 1500's, is the residence of the country's monarch. Other landmarks include the Gothic Cathedral of Notre Dame, which dates from the early 1600's, and the town hall, built in the early 1800's.

The Grand Duchess Charlotte Bridge spans the Alzette River between the historic center of Luxembourg and Kirchberg,

Industry

Luxembourg's industrial power far exceeds its small size. The nation's steel plants, which have sustained the economy for many years, are located near iron mines. These mines supply iron ore, which is essential in the production of steel. The industry is so efficient that Luxembourg ranks among the world's leading producers of crude steel.

In the 1970's, however, the value of steel production declined, and Luxembourg developed a more varied economy. High-technology industries, service industries, and international banking are now of major importance in Luxembourg.

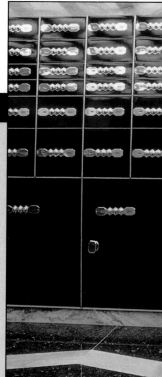

The historic section of the city of Luxembourg, *above left,* stands atop steep cliffs overlooking the Alzette and Petrusse rivers. Both charming and sophisticated, Luxembourg is an international city and an important center of the European Community.

The city of Luxembourg's small-town atmosphere is reflected in the painted signs displayed over many old shops.

A Luxembourg bank, *left,* and the stock exchange, *below,* suggest the country's importance as an international center of finance.

the city's modern section. A district of wide boulevards and new buildings, Kirchberg was constructed in 1966 to accommodate the numerous European institutions headquartered in Luxembourg, among them the European Parliament, the European Court of Justice, the European Court of Auditors, the European Investment Bank, and the European Coal and Steel Community. Because of the thousands of civil servants who work in Kirchberg, Luxembourg has earned the unofficial title of "capital of Europe."

843